AF303715

Peter Jäger

Torque wrenches - basics and calibration

Terms, basics and normative references for handling
and use of torque wrenches

____

Bibliographische Information der Deutschen Nationalbibliothek:
Die Deutsche Nationalbibliothek verzeichnet diese Publikation
In der Deutschen Nationalbibliographie; detaillierte bibliographische
Daten sind im Internet über http://dnb.dnb.de abrufbar.

ISBN: 9783756899883

Table of Contents

Table of contents

Table of contents

Table of contents

Table of contents

Foreword

This book is intended to provide concentrated information on the subject of torque wrenches.

Torque wrenches have undergone a rapid development from simple tools to precise measuring tools. This is due to advanced mechanization, ever-improving manufacturing methods and optimization processes. While it was still possible to repair a motor vehicle with a set of wrenches in the 1970s, such an "intervention" today would be a blatant safety risk and probably also the technical end of an engine, for example. Torque wrenches have also become indispensable in the medical field. Dental implants, for example, are precisely placed and fastened with the smallest torques.

Camlog-Dental-Torque wrench 0–30 N cm

Together with the general development of measurement technology, various standards and specifications have also developed around torque wrenches, which are compactly shown and explained in this book. It is intended to guide the most important standards and the reference points without the reader having to obtain, read and holistically understand these standards.

The idea for this book came from countless inquiries - by phone, in person, or by email over many years from people who were confronted with the issue and were looking for support.

Unfortunately, it can be observed that the landscape of standards and regulations - thought out and developed by experts - is often not, or only slightly, transported to the people who are supposed to implement the normative specifications. This book is intended as an introduction to the subject of torque wrenches, providing the user with basic knowledge and serving as a reference work and list of references.

Definitions

All definitions are taken from:

BIPM
International vocabulary of metrology – Basic and general concepts and associated terms (VIM)
3rd edition, 2008 version with minor corrections

This vocabulary is the reference for all metrological terms in this book.

Available on www.bipm.org.

BIPM is
The **Bureau International des Poids et Measures**. It is

- the international organization established by the Metre Convention, through which Member States act together on matters related to measurement science and measurement standards

- the home of the International System of Units (SI) and the international reference time scale (UTC).

Introduction: Metrology everywhere

Assurance of product quality is of increasing importance for every company, especially with regard to the need to maintain or consolidate its economic position on the market. Nowadays, high quality requirements for a product necessarily mean that an appropriate quality management system must be in place (keyword "product liability").

These findings are not new -modern technology and the possibilities both in mechanical manufacturing and in electronic measurement data acquisition and utilization have replaced earlier manufacturing methods. There is no longer a "fit" or a "thumb value".

The compulsion to act economically and modern manufacturing technology lead to the processualization of work.

In terms of core factors, business processes and technical or manufacturing processes hardly differ. In order to reduce vagueness, it should be noted that the further considerations and explanations refer exclusively to technical processes in distinction to services.

Physical basics

Torque

Torque (also moment of force, from Latin momentum motive force) describes the rotational effect of a force on a body.

Torque is a physical quantity in classical mechanics and corresponds to the force for rectilinear movements - but for rotational movements. A torque can accelerate the rotation of a body and bend (bending moment) or twist (torsion moment) the body.

In drive shafts, torque, together with speed, determines the transmitted power - a crucial quantity for evaluating the performance of a motor vehicle, for example.
The <u>internationally used unit</u> of measurement for torque is the newton meter.

If a force acts at right angles on a lever arm, the amount M of the torque is obtained by multiplying the force F by the length of the lever:

$$M = 1 \cdot F$$

Torque wrenches - basics and calibration

What is simply expressed by this formula can only be implemented in practice by taking numerous influences into account.

In practice - e.g. when using a torque wrench - the force does not usually act at right angles on the lever arm. In theoretical terms, the force application can be drawn into a force parallelogram. This is referred to as a pair of forces. Conversely, in statics, each torque can also be described by a force pair

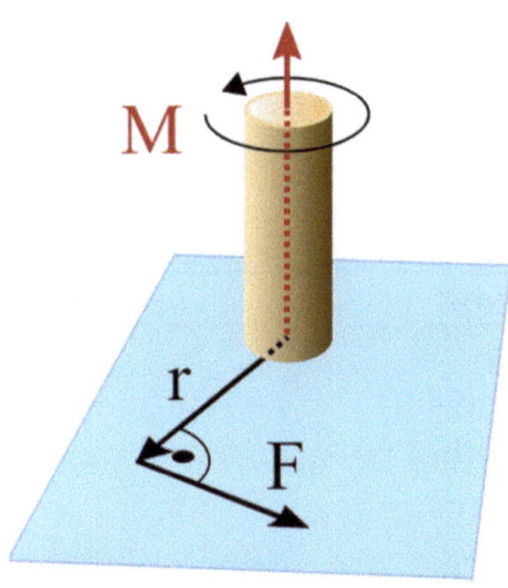

Von Pietz - Eigenes Werk, CC BY-SA 3.0,
https://commons.wikimedia.org/w/index.php?curid=10621446

Torque wrenches - basics and calibration

Newtonmeter

The Newton meter (Newton x meter) is
- the unit for the vectorial quantity torque and
- for the scalar quantities (mechanical) energy and work.

According to the formatting rules of the International System of Units, the two unit signs N and m must always be separated by a space or a multiplication point, but in practice this space is not used. The order of the factors must not be interchanged, this would mean mN = milliNewton!

In e.g. Word, "N · m" can be entered by typing "N then alt + 0183 then m".

Name of unit	Newtonmeter
Symbol	
Physical dimension	Torque, mechanical work
Formula symbol	
Dimension	
System	International sytem of units (SI)
SI unit	
Derived from	Newton, Meter

Screw- or bolted connections

A screw connection is a detachable connection of two or more parts by one or more screws. Bolted joints are the only commercially used joining technique that can be undone. All other techniques - such as welding, soldering, riveting, gluing and others cannot be undone without destruction without effort.

A bolted connection can be mentally reproduced by imagining the components between the bolt head and the nut as a tension spring. All the components that are in between - washers, sheet metal, but also loss components such as dirt, rust, unclean thread run - should be pulled together / held together by this spring.

However, it is precisely these components that act against the spring and can be assumed to be a compression spring from the point of view of these components.

By tightening the screw connection, the screw is stretched over the range of the components.

This force is referred to as the preload force. Accordingly, preload force is generated by the internal tension of the components.

$$F = \Delta l / l * E * A_s$$

With

F :	Preload force in N (in the elastic range)
Δl :	Change of length in mm
E :	Modulus of elasticity
	(Steel ~ 210.000 N/mm²)
As :	Stress cross section in mm²

VDI 2230 sheet 1 therefore prescribes: *"Bolts shall be sized to withstand the operating forces encountered and to perform the function of the joint formed."*

The yield strength - also called the yield point - indicates how far a bolt can be stretched: A cheap spring will serve as an example to illustrate the processes:

If this is stretched in the intended operating range, it returns to its original shape after the stretching stops. If this spring is stretched too far, it loses its original shape; the coiled wire stretches or breaks off. Steel behaves in a very similar way.

The yield strength is reached from the point where the metal still forms back to its original shape after being stretched. If this is no longer the case, the limit has been exceeded and the screw may no longer be used..

Influencing variables on bolted joints

A bolted connection has a wide variety of influences that must be taken into account for a strong connection:

Occurring operating forces
- Axial forces
- Transverse forces
- Bending moments
- Torsional forces
- Load change

Significant influencing variables are also introduced (around) the screw itself:
- Strength class
- Friction
- The setting behavior
- The tightening method
- Size
- Geometry of the screw.

The tightening of a bolted joint is a complex process: Tightening a bolt with the nominal torque is not yet a guarantee for correct bolting over a time span that is as unlimited as possible. The force that holds the two parts together is the preload force within the joint. The preload force stretches the bolt and is the force that prevents the connection from loosening.
The actual target value of preload force cannot be measured during assembly at present! This is only possible in individual cases (laboratory, test).

Bolt classification

Screws are subject to classification.

One finds
> 1. digit:
> This digit represents 1/100 of the minimum tensile strength.

> 2nd digit:
> This digit represents the ratio of the material yield strength to the tensile strength.

Example:
A bolt is marked 8.8

1st digit:
1/100 of the minimum tensile strength:
$8 = 800 \text{ N/mm2}$
> e.g. $8.8 = 640 \text{ N/mm}^2$

Torque and friction

There are always frictional influences during the bolting process. These are unavoidable, but must be recognized and controllable.

Reasons for friction (among others)
- Production tolerances (surface, geometry)
- Roughness and foreign particles (e.g. rust)
- Lubrication condition: dry, oiled or greased
- Material pairing and surface condition
- Molecular adhesion and temperature influence
- Coating
-

Comparison suit DM / DM and DW

Tightening with torque only:
+ Easy handling
+ Wide range of torque wrenches
- Strong frictional influence
- Large spread of preload force.

Tightening with torque and angle of rotation:
+ Higher preload force
+ Risk of bolt breakage minimized
- Hardly any frictional influence

Importance of screw fittings

Before presenting the torque wrench types, explaining their application or explaining the normative principles as well as the importance of calibration, just two concise examples will be used to show the importance that correct bolting can have:

Spiegel, 01.04.1996:
"In a crash test conducted by the trade journal Auto, Motor und Sport with the new Opel Vectra, a seat belt fastener tore off. Cause: The fastening screw at the lower left belt anchorage point "was incompletely screwed in and only bore on just under two thread turns".

Middle Hesse Blog, 06/21/2011:
After a wind turbine crash:
"That's actually impossible that this happened here. The screws should actually be able to withstand this," says Sommer, who is himself a trained mechanical engineer and knows purely technically how to assess screws."

An Internet search on the subject is revealing - time and again, connections bolted with the wrong torque are in the headlines.

Correct bolting requires that the person performing the bolting knows the type of bolting - he must have clear specifications about
- the torque to be applied
- the angle of rotation, if applicable
- the material (screw type)
have.

The design of the respective bolted joint is an engineering science; the treatment of the bolting cases would clearly go beyond the scope of this book with the target group "key users".

Typically, bolted joints are made in the range up to the yield strength.
However, there are also bolted joints that are made in the range of the yield point - i.e. in the range of the mechanical deformation of the bolt.

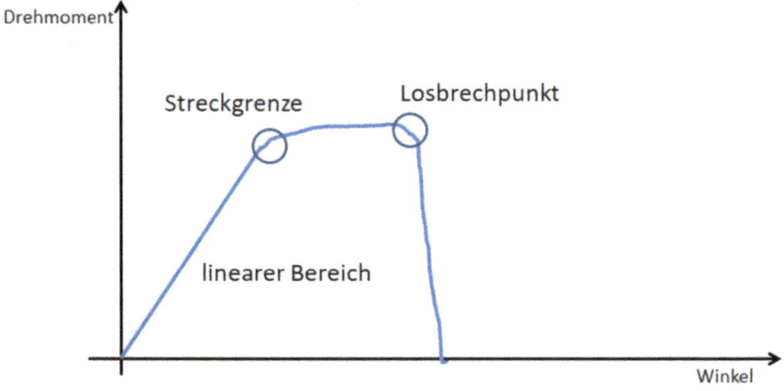

Torque/angle characteristic

Torque wrenches - basics and calibration

It is of utmost importance to know the intended type of bolting.

Since this is usually not the case in practice, it must be recognized that improper use of a torque wrench will result in bolts being screwed into the wrong area - thus creating the possibility of loosening as well as overstretching and destroying the bolt.

The torque wrench user must therefore strictly adhere to the specifications and not deviate from them "by experience" and "with arm feeling".

Torque wrenches – types and styles

Torque wrenches are available in different
- mechanical or
- electronic versions

and different operating principles.

Mechanical torque wrenches:
- Kink wrenches, kink when the desired torque is reached, thus preventing further initiation of a torque.
- releasing torque wrenches (click wrenches).
- Slip wrenches.

Click torque wrenches

Mechanical release wrenches or click wrenches are the most common type of torque wrench. But it is precisely this type of wrench that requires expert handling by the user - this is where the most frequent operating errors are observed.

The operator must immediately stop the initiation of torque after the trigger signal or "click".

The click wrenches get their name from the corresponding sound that is heard when the set torque is reached.

Torque wrenches - basics and calibration

This type of torque wrench is widely used in both hobby and professional workshops.

Von Torquemaster, CC BY-SA 3.0,
https://commons.wikimedia.org/w/index.php?curid=18491897

On a releasing type torque wrench, a specific target torque is set by means of a scale or with the aid of a testing device (this process is often confused with calibration or adjustment).

As soon as the torque wrench reaches the preset torque at the drive part (force axis), this is indicated by an audible, tactile and/or visible signal, usually by a perceptible click/crack or a colored indicator popping out.
They often consist of a steel tube, the length of which is adapted to the range to be covered (compare N.m !), with a release mechanism as well as an adjusting device connected to the handle, including a scale. The tool carrier is usually a reversible or reversible ratchet (also called a ratchet).

Torque wrenches - basics and calibration

Force is then applied to the bolting point via an inch square typically 1/4", 3/8", 1/2", 3/4", 1" or 1 1/2" on sockets. Other common tool holders include the rectangular holder or dovetail for interchangeable tool heads in various profiles such as open-end, ring, open-ring or hex wrenches, among others.

The world's first self-releasing torque wrench was patented and launched in 1938 by Saltus-Werk Max Forst, Solingen.

Slipper torque wrenches

Also called "slipping wrenches", slip-through torque wrenches are the most convenient manual way to use the correct torque to tighten bolts. They work with a slip clutch and interrupt the power transmission as soon as the selected torque is reached. This eliminates the possibility of overtightening the bolted joint. Slip torque wrenches are comparatively expensive.

Mechatronic torque wrenches

The torque insertion is similar to the triggering torque wrench, in addition the torque is measured digitally (click and final torque) as with the electronic torque wrench. It is therefore a combination of electronic and mechanical measurement.

Buckling torque wrench

The buckling torque wrench or "buckling wrench" buckles when the set torque is reached at the pivot point; the application of force is thus interrupted immediately.

Compared to a click wrench, it offers the advantage that the mechanical interruption means that leverage is no longer achieved even after the wrench has been released.

Beam torque wrenches

Indicating torque wrenches - i.e. directly measuring torque wrenches such as Beam-type torque wrenches indicate the torque reached during the bolting process. The principle of operation is usually quite simple. In the beam-type wrench, the upper thinner indicator rod remains straight; the main shaft together with the attached scale bends in proportion to the applied force.

Indicating torque wrenches use a mechanical scale, dial indicator, or electronic display to indicate / display the value of torque applied by the tool to the output member. Many designs use solid levers with a torsion bar as the drive, whose rotation relative to a scale coupled to the lever can be read as a measure of torque.

Torque wrenches - basics and calibration

The simplest form of this type consists of a long lever arm between the handle and the head of the wrench,

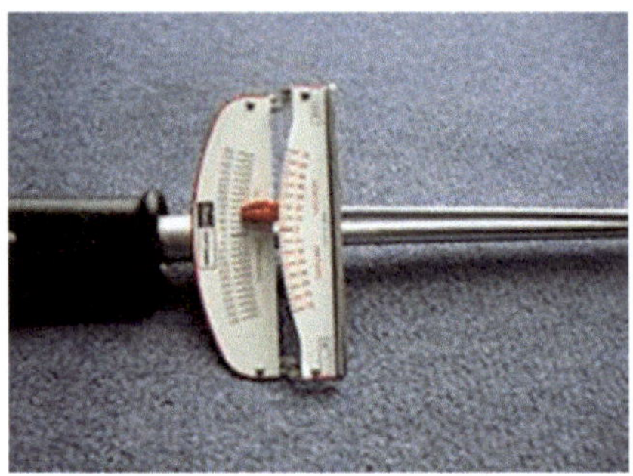

Indicating torque wrench
Von EncMstr - Eigenes Werk, CC BY-SA 3.0,
https://commons.wikimedia.org/w/index.php?curid=658567

made of a material that bends under the tightening torque. The second rod with an indicator is fixed at one end and runs parallel to the lever arm.

This part remains without load and does not twist. A scale is attached to the end of the handle. The bending of the main arm when a torque is generated causes a displacement of both arms, which can be read on the scale. When the desired torque is indicated, the operator must stop applying force.

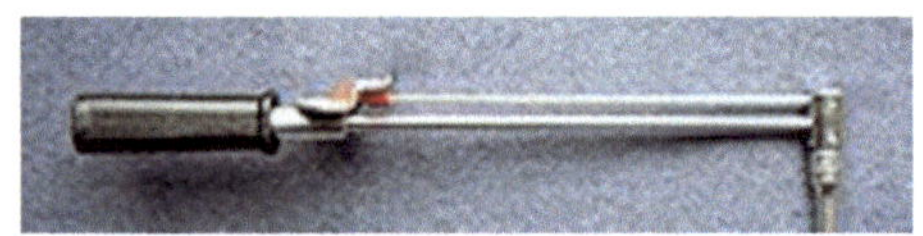

Von EncMstr - Eigenes Werk, CC BY-SA 3.0,
https://commons.wikimedia.org/w/index.php?curid=658567

The Chrysler torque wrench

The principle of today's flat bar wrench was originally invented by Walter P. Chrysler in the 1930s.

He had conceived the principle for his Chrysler Corporation; it was further developed by Micromatic Hone Corporation.
This type of torque wrench was eventually manufactured and sold by the Cedar Rapids Engineering Company.

Their regional sales manager, Paul Allen Sturtevant of Elmhurst (IL), further developed Chrysler's imperfect idea, patented his own invention in 1938, and marketed it through his own P. A. Sturtevant Company, Addison (IL).

Torque wrenches - basics and calibration

Electronic torque wrench

Electronic (digitally displaying) torque wrenches carry out the measurement usually by a strain gauge array (DMS) on the torsion bar. Typically, the strain gauge is connected as a Wheatstone measuring bridge, which is applied once or several times.
The bridge signal is converted into the unit to be displayed (N-m, lbf-ft, etc.) by means of a measuring amplifier and an arithmetic unit and shown on the display.

Depending on the design and quality, several different bolting cases (bolting data or limit values) can be stored. These entered default values are permanently displayed during the tightening process.

Today's wrenches also offer the option of initially specifying the torque target value and an angle extension. After reaching the torque target value (which can be read on the display and / or signal tone or, if necessary, handle vibration), a torque / angle wrench can automatically switch to an angle display so that the bolting operation can be completed according to the specifications.

Electronic torque wrenches usually store the values of all measurements performed in an internal measured value memory. This can be read out via interface (wired or wireless) or printed directly via a connected printer.

Torque wrenches - basics and calibration

With electronic torque tools, safe handling by the user is required. Although the optical or acoustic device of the wrench signals that the tightening torque has been reached - the strict reaction to the signal and the adjustment of the application of force is up to the user.

In order to ensure the quality of the bolted joint, these bolting operations require a high level of triggering accuracy when the set torque is reached, with a corresponding reproducibility of the values (repeatability).

This requires a clearly recognizable trigger signal and the least possible influence of handling differences, for example due to fast or slow tightening.

Programmable electronic torque wrenches

A programmable electronic torque wrench works like an electronic torque wrench; in addition, an angle is determined from a joining torque/threshold value. The angle is measured with the aid of an angle sensor or electronic gyroscope. Based on the angle measurement, already screwed connections can be detected. Statistical evaluations can be carried out using the internal measured value memory. Curve progressions can be analyzed by means of software via the integrated curve progression memory (force-path diagram). This type of torque wrench can also determine the breakaway torque, continued torque and final torque of a bolt course. Using a special measuring method, even the yield point can be displayed (yield point controlled tightening).

This torque wrench is mainly used by automotive manufacturers for the documentation of torque-angle controlled tightening.

In 1995, Saltus-Werk Max Forst GmbH applied for an international patent for the first electronic torque wrench with reference-armless angle measurement.

Torque wrenches - basics and calibration

Reference torque- / transfer wrenches

Reference wrenches or transfer wrenches are not screwdriving tools but purely laboratory equipment. They consist of a precision measuring disk and a lever arm of a length corresponding to the range.
Reference wrenches are used for the calibration of torque (wrench) calibration devices.

"Transfer torque wrenches are special torque measuring devices which, by their design, enable the torque to be introduced via a lever arm (comparable to the design of torque wrenches) and are insensitive to superimposed transverse forces and bending moments in accordance with the required measurement uncertainty. Due to their special design, transfer torque wrenches enable the calibration of torque wrench calibration equipment taking into account the actual force application conditions with the simultaneous possibility of varying the force application parameters according to the variation range of the transverse forces and bending moments occurring during the calibration of torque wrenches." (compare / from DAkkS-DKD-R-3-7)

Torque wrench handling

Torque wrench application

Success for a proper bolted joint lies in knowing the torque specifications of the bolts to be used or better, the bolt case.

Torque values for most fasteners are based on clean, dry and undamaged threads.
The load on the screw depends on the friction created by the threads as the screw is tightened. If the threads are oiled or lubricated, this will reduce friction and increase the load applied to the bolt. This can overload the bolt. There is a risk that the bolt will be stretched or broken.

When tightening a bolt or nut with an ordinary wrench, the fastener is first tightened to the point where it is snug, but not too tight. Then the bolt case is tightened with a torque wrench to the given specifications on.

Torque wrenches - basics and calibration

Handling of click torque wrenches

When dealing with triggering torque wrenches, some things must be observed.

- The torque is always set from the low value to the high value.

- The lock must be seated, the wrench must not be able to adjust.

Torque wrench application rules

As with any measuring instrument, there are operating and usage guidelines for handling torque wrenches. Basically and for all wrench types, the following 5 simple rules of thumb apply:

Torque wrench rules

1. Never loosen bolted connections with a torque wrench. A torque wrench is a sensitive and precise tightening tool only.

2. Treat the torque wrench like a sensitive measuring instrument. A torque wrench is a precise measuring tool!

3. Use one hand only! A torque wrench must be guided with a smooth and continuous movement until it triggers / reaches the target value.

4. Do not use an extension - apply the force at the correct point! An extension prevents the set value from being transmitted correctly. Many wrenches have a marking on the handle - here (and only here!) the force should be initiated!

5. Release the torque wrench! After finishing the work, the torque wrench must be released - i.e. reset to the lowest value in order to release the spring tension.)

Torque wrenches - basics and calibration

A torque wrench can prevent over-tightening, but it cannot prevent it!

An application error often seen when using triggering wrenches is tightening further "for safety". This tightening could have been done without a torque wrench - it is technically worthless. When the selected torque is reached and the wrench triggers, the tightening process is complete!

For wrenches that emit a signal tone instead of the mechanical release, this procedure applies analogously. If a signal sounds already at the beginning of the screwing curtain without having moved the torque wrench, the screw connection is already too tight. In this case, a new start must be made: the screw must be loosened again; the correct torque is applied with a torque wrench.

After finishing the (day's) work these few actions must become a daily routine:

-    Release torque wrench

-    Store torque wrench properly: The wrench should always be stored protected in a box and protected from shocks.

Torque wrenches - basics and calibration

Torque wrench don'ts and dos

The following photo series is intended to visualize the most common handling errors.

All photos are posed, the handling may be exaggerated.

Never loosen screw connections with a torque wrench! (if you do: turn in for calibration!)

Torque wrenches - basics and calibration

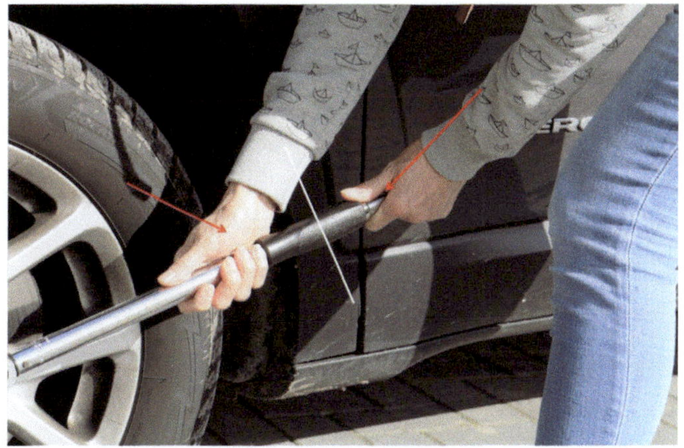

Wrong: two-handed force application "somewhere" -
not in the middle of the grip

Wrong: force application too far back (here: the
moment setting is used as an additional lever)

Never use feet or extensions if your own strength is
not sufficient!

This is how it works: Apply force vertically;
pull evenly!

Consequences of improper handling

A bad torque wrench or the wrong handling of a good torque wrench can have nasty consequences. Excessive torques overload the material, so that bolts either crack immediately or wear out more quickly due to material fatigue. The threads are often damaged and, in the worst case, the bolt gives way while driving.

But even if the screw holds, too much torque can make it impossible to get it off. In such cases, the effort and costs involved usually increase rapidly. Too little torque is also a serious source of danger. If, for example, the wheel bolts come loose because they were not tightened properly, the carelessness can quickly become life-threatening.

Normative references

In December 1989, the Product Liability Act (Gesetz über die Haftung für fehlerhafte Produkte - ProdHaftG) was enacted in Germany; similar laws have been released in other countries in order to protect customer rights.

It regulates the liability of a manufacturer for defective products. The law aims to compel the manufacturer to exercise the utmost care in the manufacture of products in order to minimize or eliminate hazards that may emanate from a product and thus protect the consumer. While the law regulates the conditions of liability, supplementary regulations and standards provide enforcement and implementation guidelines.

One of these specifications is the ISO 9000 series of standards, according to which many (not only manufacturing) companies are certified today. It specifies requirements for quality management systems that are intended to serve as tools for achieving defined quality standards and thus for product safety and protection of the end user.

Certification to ISO 9000 is proof that the duty of care required by the Product Liability Act has been met by establishing suitable processes and setting up an integrated quality management system.

Torque wrenches - basics and calibration

The ISO 9000 series requires that measuring and testing devices – this includes torque wrenches - are subject to monitoring and - depending on their use - are also calibrated.

This requirement, which is formulated in ISO 9001, is specified in various standards and regulations with regard to practical implementation.

The most important references are listed below. The list is not exhaustive.

- DKD 4 Traceability of measuring and test equipment to national standards
- International Vocabulary of Metrology Basic and General Concepts and Associated Terms (VIM) www.bipm.org
- ISO/IEC 17000:2005-03: Conformity assessment - terms and general principles
- ISO 10012:2003 Measurement management systems - Requirements for measurement processes and measuring equipment

- ILAC-G8:03/2009: Guidelines on the Reporting of Compliance with Specification
- DIN EN ISO 14253-1:1999: Geometrical product specifications (GPS) - Inspection of workpieces and measuring instruments by measurement; Part 1: Decision rules for determining conformity or non-conformity with specifications
- UKAS M3003: The Expression of Uncertainty and Confidence in Measurement (Edition 2, 2007), Appendix M Assessment of Compliance with Specification

The three most significant standards will now be addressed in parts, relevant passages will be discussed:

- ➤ DIN EN ISO 9001:2015 Qualitymanagementsystems – Requirements
- ➤ IATF 16949 Requirements for quality management systems for series and spare parts production in the automotive industry (BMW, Chrysler, Daimler, Fiat, Ford, General Motors, PSA, Renault, VW)
- ➤ DIN EN ISO 17025:2018 General requirements for the competence of testing and calibration laboratories

ISO 9001:2015

For test- and measuring equipment, an essential section of ISO 9001:2015 is section 7.1.5 ff. It is the binding basis for the establishment and operation of a measuring equipment management system:

ISO 9001:2015, Ziff 7.1.5 ff

The organization shall identify and provide the resources needed to ensure valid monitoring and measurement results to demonstrate conformity of products and services to specified requirements.

The organization shall ensure that the resources provided are
a) Are appropriate for the particular type of monitoring and measurement activities undertaken;
(b) are maintained to ensure their continuing suitability.

The organization shall retain appropriate documented information as evidence of the suitability of the resources for monitoring and measurement.

This provides a fixed framework for the deployment and use of measuring and testing equipment in a (certified) company.

In the following paragraph, it gets very specific:

„7.1.5.2 Metrological traceability
When metrological traceability is a requirement,... „

In the above-mentioned introductory sentence, the standard makes already a restriction with the word "if" - obviously not every measuring and test device is affected by the measures listed below.

This means that when setting up a measuring equipment management system, the measuring equipment in operation should be considered not only with regard to its function but also with regard to its application.

... or is considered by the organization to make a significant contribution to establishing confidence in the validity of the measurement results, the measuring equipment shall:

> a. *be calibrated, verified, or both, against standards traceable to international or national standards at specified intervals or prior to use; if no such standards exist, the basis for calibration or verification shall be retained as documented information;*

In addition to the technically oriented requirement to maintain and use precise measuring and testing equipment, this sentence also contains an approach oriented towards sustainability: it is therefore not sufficient to procure calibrated measuring and testing equipment or to have it calibrated once, but this measure must be repeated at intervals or on a case-by-case basis.

In this (current) version of ISO 9001, traceable calibration is even required. This is often interpreted to mean that every measuring device must be traceably calibrated. However, the exact wording is:

"... be calibrated, verified, or both, against standards traceable to international or national standards, ..."

This certainly allows the use of a factory / working standard on which the utility measuring equipment is calibrated. Example: a calibration system for torque wrenches is traceably calibrated - in Germany by a DAkkS calibration - and is available in the company (= "the organization"). Thus, depending on the application, the operational torque wrenches can be calibrated as factory calibration.

The authorized person or measuring device user is specifically urged in the further text to also visually mark his devices and to take protective measures:

The measuring equipment must ...

   b.   be labeled in order to be able to determine the status

Torque wrenches - basics and calibration

Examples may be:

A custimized, company own calibration sticker:

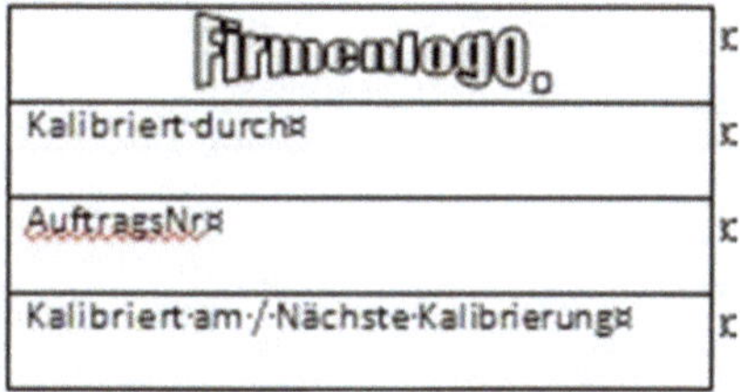

The three lines contain
- name of the technician or performing service
- order No (for easy identification in ERP)
- date of calibration OR due date

Or, in the case that a device is not calibrated (and thus may not be used in a quality-relevant manner) a sticker such as:

An auditor - but much more importantly in everyday operation, the user - recognizes the status of the measuring device at a glance.

Torque wrenches - basics and calibration

Whether the decision (e.g. "not calibrated") is correct can be debated. However, this takes an audit to another level.

Important at this point is: the auditor recognizes that all measuring equipment has been taken into account and decisions have been made regarding its calibration status.

The standard further specifies:

The measuring device must ...

be protected from setting changes, damage or deterioration, which would invalidate the calibration status and consequently the measurement results.

Here, again the holder of the measuring device has a direct obligation:

He must take suitable measures to prevent (at least) unintentional manipulation of the measuring device.

Torque wrenches - basics and calibration

ISO 9001 further states:

The organization shall determine whether the validity of previous measurement results has been affected if the measuring equipment is found to be unsuitable for its intended use, whereupon the organization shall take appropriate action as necessary.

A look into the (recent) past is required with regard to the use of the measuring device:
- Where / on which object was the measuring device used for measurement?
- does this have quality-relevant effects?
- Does this even have safety-relevant effects (aviation or automotive safety!)?
- What must be done if one of the last two questions is answered in the affirmative? Recall? Customer information? Re-setting of a production line?

ISO/IEC 10012:2004-03

ISO/IEC 10012:2014 is the normative basis for measurement management systems.

A measurement equipment management system ("measurement management system") should also ensure that measurement equipment and processes are suitable for the intended use, that possibly incorrect measurement results are detected at an early stage, and that the risk and effects can be made controllable.

In the following section, sections are shown which directly affect measuring equipment holders / users. This group of persons is already addressed in the foreword of the standard:

Reference may be made to this international standard:
- *By a customer when specifying the characteristics of the products required;*
- *by a supplier in defining the characteristics of the products offered;*
- *by legislative bodies or regulatory authorities; and*
- *in evaluations and audits of measurement management system*

This standard does, however, provide some significant requirements that directly affect the measuring equipment holder / user:

7.1 Metrological confirmation
    7.1.1 General
    Metrological confirmation (... shall be designed and implemented to ensure that the metrological characteristics of the measuring equipment meet the metrological requirements for the measurement process. The metrological confirmation includes the calibration and verification of measuring equipment

A clear obligation to (regularly) perform calibrations is expressed.

The importance of specifying measurement uncertainties, without which it is not possible to speak of a technically correct and complete calibration, is emphasized in the further text as a means of evaluation.

Even for the determination of the calibration intervals specifications are made:

7.1.2 Intervals of metrological confirmation

The methods used to establish or modify intervals between metrological confirmations shall be described in documented procedures. These intervals shall be reviewed and adjusted as necessary to ensure continued compliance with the specified metrological requirements.

...

Each time a faulty measuring device is repaired, adjusted or modified, the interval for metrological confirmation must be evaluated.

See also chapter "Determination and adjustment of calibration intervals" in the second part of this book.

ISO 6789 – 1

Assembly tools for screws and nuts –
Hand torque tools –
Part 1: Requirements and methods for design conformance testing and quality conformance testing: minimum requirements for declaration of conformance (ISO 6789-1:2017)

Former issue ISO 6789:2003 has been modified / updated and split into two parts. This document specifies requirements for design and manufacture, including the content of a declaration of conformity.

ISO 6789 – 2

Assembly tools for screws and nuts – Hand torque tools –
Part 2: Requirements for calibration and determination of measurement uncertainty
ISO 6789-2 specifies the requirements for traceable calibrations. It contains a procedure for the calculation of measurement uncertainties and proposes a torque measurement device calibration procedure for use in the calibration of hand-operated torque screwdriving tools.

DAkkS-DKD-R 3-7

This guideline applies to calibrations of:
- Transfer torque wrenches used as transfer standards for the calibration of torque wrench calibration equipment,
- Torque wrenches of higher accuracy than classified according to DIN ISO 6789.
-

Described is a procedure for classification as well as for determining the relative expanded uncertainty of measurement of these devices.
The definition of an indicating torque wrench is important:

- indicating torque wrench: the entire device from the tool holder with insertion tool through the torque transducer up to and including the indicating device.

This means that this entire chain must also be presented for calibration in order to obtain a truly comprehensive calibration.

ISO 6789-2:2017 requirements

With the amendment of DIN EN ISO 6789, the effort for a calibration has approximately tripled. Accordingly, the processing time has almost tripled, and the costs have increased significantly.
This even led to calibration service providers withdrawing from the market because it was no longer possible to work economically.
The calibration laboratories that still offer calibrations for torque wrenches often have long lead times.

The reason for the tripling of the calibration effort are the different influencing variables on a torque wrench, as listed before.

The DIN EN ISO 6789:2017 has - new - two parts. While the first part is largely identical in content to the previous version, part 2 differentiates between possible error influences and attempts to systematically record these by means of numerous series of measurements and to determine characteristic values - so-called b-parameters - and incorporate them in the evaluation of the torque wrench.

The aim is to establish a comprehensive measurement uncertainty consideration - accordingly, the regulation states: "Determination of type B measurement uncertainties caused by the torque screwing tool".

Variation due to the reproducibility of the torque tool b_{rep}

ISO 6789-2017, No. 6.2.2:

Reproducibility is affected by the ability to identify exactly the value at which loading should be stopped
for indicating torque tools Type I and the ability of the mechanism to return in exactly the same place
each time after adjustment of the tool in the case of setting torque tools Type II. For both Type I and
Type II tools, it includes parallax errors.

For torque tools of all types, the following procedure for determining the comparison precision b_{rep} is described. The tool shall be loaded with the load sequence described in ISO 6789-1:2017, 6.5 exclusively with the lowest specified torque value and the values shall be recorded.

The sequence must be performed four times and the torque screwdriver must be removed from the calibration system between each sequence."

For the calibration station, this means having to take 10 readings for each position of the square - i.e. a total of 40 measurements.

Variation due to geometric effects of the output drive of the torque too b_{od}
ISO 6789-2017, No. 6.2.3.2:

Ratchets, hexagon and square drive outputs of the torque tool in particular have an influence since
they can potentially run out of true and if not used in the same orientation each time, they can cause
variation of reading. Interchangeable drive ends can also cause variation.
Interchangeable drive ends of the torque tool including the centre distance shall be identified and
documented.

Torque wrenches - basics and calibration

The method described below is used to determine the variation bod caused by the output part. ... .

If the output part cannot be rotated, this variation must be set to zero.

The tool shall be positioned on the calibration system according to ISO 6789-1:2017, 6.5, and subjected to five preloadings at the lower limit value of the measurement range, T_{min}.

The torque wrench is removed from the calibration system and the output part is rotated 60° (for hexagonal output parts) or 90° (for square output parts). For at least four positions evenly distributed over 360°, ten measurements each are recorded at the lower limit of the measuring range T_{min} without changing the load application point.

For the calibration technician, this means having to record 10 measured values per position of the square - i.e. a total of 40 further measurements.

Variation due to geometric effects of the interface between the output drive of the torque tool and the calibration system b_{int}

ISO 6789-2017, No. 6.2.3.3

Hexagon and square drive interfaces between the output drive of the torque tool and the calibration system have an influence since they can potentially run out of true and if not used in the same orientation each time, they can cause variation of reading.

The interface between the output drive of the torque tool and the calibration system shall be identified and documented.

The following method is described for the determination of the variation b_{int} due to the drive interface.

This value may be determined statistically for a sufficient number of specimens (at least 10) of a model of tool and its determination does not need to be repeated each time for future calibrations of this model.

The tool shall be positioned on the calibration system according to ISO 6789-1:2017, 6.5, and subjected to five preloadings at the lower limit value of the measurement range, T_{min}.

The torque tool is removed from the calibration system and the drive interface is rotated by 60° (hexagonal drive output) or 90° (square drive output). Ten measurements are recorded for each of at least four positions distributed evenly over 360°, at the lower limit value of the measurement range, T_{min}, without changing the load application point.

The calibration technician needs to take 10 measurements for each position of the square - i.e. another 40 measurements.

Variation due to the variation of the force loading point, b_l

ISO 6789-2017, No. 6.2.4

Most torque wrenches have some variation in torque observed depending on the exact force loading point on the handle. This does apply to both indicating and setting wrenches, but not to torque screwdrivers of either type. For torque screwdrivers, the value of bl shall be set to zero.

Where the loading point is not marked on the torque tool and no manufacturer information is available, the dimension from the axis of rotation to the loading point used shall be documented.

The following method is described for the determination of the force loading point variation, b_l. This value may be determined statistically for a sufficient number of specimens (at least 10) of a model of tool and its determination does not need to be repeated each time for future calibrations of this model.

The tool shall be positioned on the calibration system according to ISO 6789-1:2017, 6.5, and subjected to five preloadings at the lower limit value of the measurement range, T_{min}.

Ten measurements are then recorded for each of two positions with changed force loading point, at the lower limit value of the measurement range, T_{min}. The two force loading points shall be 10 mm on either side of the centre of the hand hold position or the marked loading point.

Remark / Evaluation
ISO 6789-2:2017 has the following addition in all sections on b-parameters:

This value may be determined statistically for a sufficient number of specimen (at least 10) of a model of tool and its determination does not need to be repeated each time for future calibrations of this model..

Torque wrenches - basics and calibration

This requirement (measurement of at least 10 torque wrenches of one type) can of course be met very quickly by the torque wrench manufacturers who offer calibrations. Independent laboratories, which randomly test the most diverse torque wrenches of the most diverse are clearly at a disadvantage here.

If one lists all the measurements required for the comprehensive calibration of a torque wrench in both directions, one obtains the following overview:

Measuring series	No. Of Measurements
Calibration cw	15
Calibration ccw	15
Determination b_{rep} cw	40
Determination b_{od} cw	40
Determination b_{int} cw	40
Determination b_l cw	20
Determination b_{rep} ccw	40
Determination b_{od} ccw	40
Determination b_{int} ccw	40
Determination b_l ccw	20
Summe:	**310**

Cw = clockwise = right
Ccw= counterclockwise = left

Instead of 30 measurements for as before, up to 310 measurements must be taken for a complete calibration procedure.
There are indications from the Torque Technical Committee that these b-values must be determined again analogous to the calibration interval, i.e. every 12 months. This leads to uncertainty among the calibration laboratories - especially among those who want to implement the specifications of the DKD and the technical committees in a serious and complete manner.

Conclusion: In its current form, ISO does not stipulate that the b-parameters have to be determined again at 12-month intervals.

However, in view of the large number and variety of torque wrench types, it is generally to be expected that a non-manufacturer calibration laboratory will have to carry out these measurements with the corresponding effort.

Calibration

Why calibration?

All test- and measurement equipment is exposed to influences during use and also during storage which can permanently change their metrological properties. This is particularly true for torque wrenches:

- the internal lubricants can harden
- the spring can fatigue
- the internal release mechanism wears out
- overload
- improper use
- influences in the working environment (simply drop torque wrench – happens!)
- failure / malfunction of the electronics

can have striking negative metrological influences. Calibration is used to determine whether measuring and testing devices have the required accuracy.

Definition Calibration:

The International Dictionary of Metrology defines calibrations as follows:

"Activity which, under specified conditions, in a first step establishes a relationship between the quantity values provided by standards with their measurement uncertainties and the corresponding indications with their associated measurement uncertainties, and in a second step uses this information to establish a relationship by means of which a measurement result is obtained from an indication".

NOTE 1 The result of a calibration may be expressed in terms of an indication, a calibration function, a calibration diagram, a calibration curve, or a calibration table. In some cases, it may consist of an additive or multiplicative correction of the indication with the associated measurement uncertainty.

NOTE 2 Calibration should not be confused with adjustment of a measurement system, which is often erroneously called "self-calibration", nor with verification of calibration.

NOTE 3 Often, only the first step in the definition above definition is considered as calibration.

Calibration is therefore a measurement process for the reliably reproducible determination and documentation of the deviation of a measuring instrument or a material measure from another instrument or another material measure (standard).

According to the definition of the VIM of JCGM 2008, a second step is mandatory for the definition of calibration:

The consideration of the determined deviation in the subsequent use of the measuring instrument for the correction of the read values. These definitions, which are nevertheless very theoretically formulated, will be explained in the following:

Like any measuring device/device, torque wrenches are also characterized by three basic parameters; these are parameters of any calibration:

- ✓ Precision:
  How accurate is the displayed value?

- ✓ Repeatability:
  Can this value be repeated x times for x measurements?

- ✓ Linearity:
  An adjustable torque wrench with a range of 0 ... 100 N.m will. If values such as 60 - 80 -

100 N.m are set - does the trigger value always have the same deviation?

Example: if triggered at 58 - 78 - 98 N.m, the key has a deviation of - 2. This deviation can be easily corrected: either by a correction table or by an adjustment.

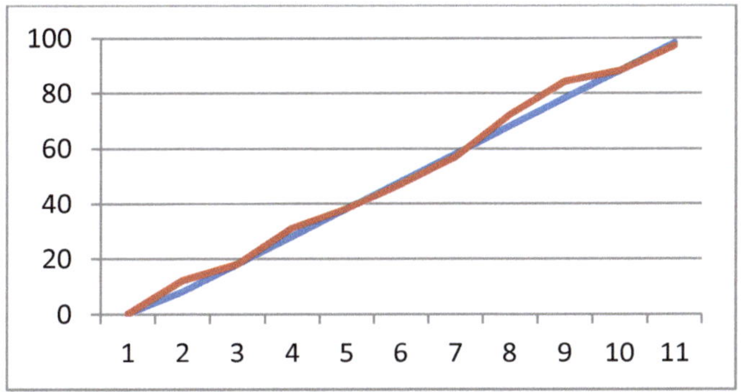

However, if the wrench were to trip at 58 - 63 - 100, for example, the tripping value would "eject" around the set torque value - the wrench would be non-linear and much more critical to handle - and definitely of poorer quality (red example line).

The mentioned characteristics are to be clarified at the following graphic:

1. the measurement result scatters in a wide range - the torque wrench is useless for resilient measurements.

1. the result is not very precise, but repeatable in a certain (wide) range - the wrench can be used for specific tasks.

2. the scatter is low (= good repeatability), but has a deviation from nominal values. The wrench is good, but needs correction (spreadsheet or software).

3. good, precise result with high repeatability and small linearity error.

Torque wrenches - basics and calibration

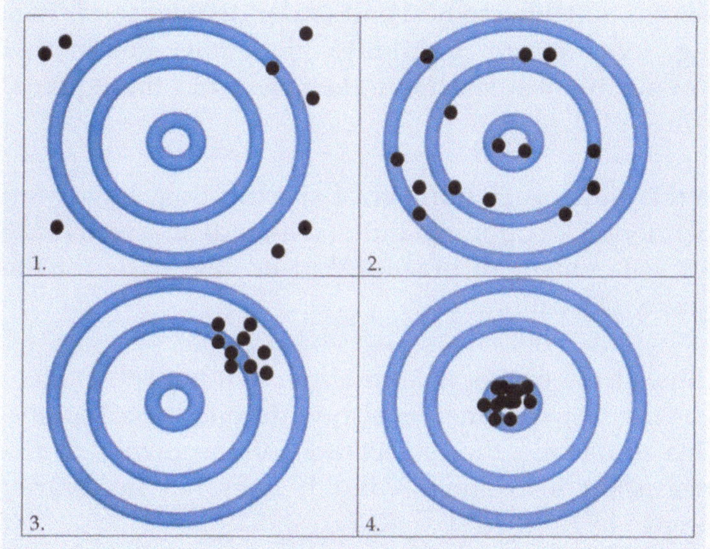

With a calibration, the relationship between the output values of a measuring instrument or a measuring device or the values represented by a material measure or a dimensional scale and the associated values of a measurand determined by standards is now determined under specified conditions.

This means: there is an (accurate) standard (which of course must also be calibrated in a traceable manner) and a (inferior) measuring device.

Using a suitable method and under defined conditions (e.g. temperature, humidity), it is now determined how far the test specimen deviates from the nominal value.

In this sentence, a number of specifications have been mentioned - sounding quite self-evident - which are not self-evident in practice, but are the basis for the quality of a calibration.

Thus, at the end of a calibration, it is not self-evident that the test specimen is also within its specifications. This is usually the expectation of the owner of the measuring instrument - but it does not necessarily have to be fulfilled.

A calibration certificate initially only documents the (metrological) properties of the test specimen found at the time of calibration at an accredited calibration laboratory on the basis of a normalized / standardized procedure with a validated measurement method and provided with an uncertainty factor.

Thus, the International Dictionary of Metrology defines the term "calibration" as follows:

"Activity which, under specified conditions, in a first step establishes a relationship between the quantity values provided by standards with their measurement uncertainties and the corresponding indications with their associated measurement uncertainties, and in a second step uses this information to establish a relationship by means of which a measurement result is obtained from an indication."

Only if also - as possible - a conformity statement is required, the expectation of the client is met, provided that the measuring instrument is within the specifications - a "simple calibration" does not include this.

Accreditation bodies and -laboratories

DAkkS

DAkkS is the German language abbreviation for "German Accreditation Body". This body is the national accreditation body of the Federal Republic of Germany. It acts according to the Accreditation Body Act (AkkStelleG) as the sole service provider for accreditation in Germany.

The DAkkS does not operate for profit. It is a limited liability company (GmbH) whose shareholders are the Federal Republic of Germany, the Federal States of Bavaria, Hamburg, Lower Saxony, North Rhine-Westphalia and Saxony-Anhalt and, through the Federation of German Industries (BDI), German industry.

Among other things, the DAkkS accredits calibration laboratories, i.e. a calibration laboratory is "approved" after a (strict) examination of the fulfillment of specified requirements. It has proven that it works according to the DAkkS guidelines and is allowed to use the DAkkS logo in its calibration certificates.

A2LA

American Association for Laboratory Accreditation (A2LA) is among the largest accreditation bodies in the world and the only independent, 501(c)3, non-profit, internationally recognized accreditation body in the United States that offers a full range of comprehensive conformity assessment accreditation services.

Established in 1978 as a public service membership society, A2LA is dedicated to the formal recognition of competent testing and calibration laboratories (including medical laboratories), biobanking facilities, inspection bodies, product certification bodies, proficiency testing providers, and reference material producers. A2LA has over 4,000 actively accredited certificates representing all 50 US states and more than 50 countries.

(taken from the A2LA Homepage a2la.org)

CNAS

China National Accreditation Service for Conformity Assessment (hereinafter referred to as CNAS) is the national accreditation body of China unitarily responsible for the accreditation of certification bodies, laboratories and inspection bodies, which is established under the approval of the Certification and Accreditation Administration of the People's Republic of China (CNCA) and authorized by CNCA in accordance with the Regulations of the People's Republic of China on Certification and Accreditation.

CNAS was founded on March 31, 2006 by merging the former China National Accreditation Board for Certifiers (CNAB) and China National Accreditation Board for Laboratories (CNAL).

The purpose of CNAS is to promote conformity assessment bodies to strengthen their development in accordance with the requirements of applicable standards and specifications, and to facilitate the conformity assessment bodies to effectively provide service to the society by means of impartial conduct, scientific means and accurate results.

(taken from the CNAS Homepage cnas.org.cn)

Calibration laboratory with accreditation

When talking about an accredited calibration (or testing-) laboratory, it is usually meant an accreditation by an accreditation body and based on ISO 17025, "General requirements for the competence of testing and calibration laboratories".

ISO/IEC 17025:2018 states – new and in contrast to previous issues:

"This document has been developed with the aim of promoting confidence in the work of laboratories. This document contains requirements for laboratories to demonstrate that they are competent and capable of producing valid results. Laboratories that comply with this document will also operate in general accordance with the principles of ISO 9001."

Whereas in the earlier document the message was "accreditation to ISO 17025" does not necessarily have anything in common with "certification to ISO 9001", the standards committees have reacted to reality and recognized
- That the vast majority of accredited calibration laboratories also have certification to ISO 9001
- That there are numerous overlaps in content - here: the general QM parts - in the two quality management manuals

These two segments are not contradictory, but rather complement each other, and the handling practiced to date means a duplication of work and a double burden for all parties.

Accordingly, with the edition of 17025:2018, a common general QM part can / should be created and used for certification and accreditation.

Nevertheless, the following applies: if a suitable calibration laboratory is sought (out), the focus should accordingly not be on certification according to ISO 9001, but on accreditation according 17025.
Whereas ISO 9001 makes fundamental demands on quality management, 17025 is specifically tailored to calibration- and testing laboratories:

In this regulation, the technical requirements for a calibration laboratory alone include a catalog of measures for each:
- The personnel
- The premises and environmental conditions
- The test and calibration procedures and their validation
- The selection of procedures
- Procedures developed by the laboratory

- Procedures not specified in normative documents
- Validation of procedures
- Estimation of measurement uncertainty
- Steering of data
- The facilities
- The metrological traceability
- Special requirements
- Reference standards and reference materials
- Sampling
- Handling of test and calibration items
- Assurance of the quality of test and calibration results

A calibration laboratory which has been accredited (with appropriate effort) according to ISO 17025 can always be selected as a trustworthy calibration laboratory for the accredited parameters.

Calibrations performed in an accredited laboratory stand for the reliability of the measurement results (including uncertainties) and are indispensable for certification according to EN ISO 9001, which requires traceability to national standards for the measuring equipment used.

However, calibration services without any accreditation are not recommended – international metrological institutes such as NIST (USA) have strict definitions for traceability.

Despite the distinction from ISO 9001 described above, an accredited calibration laboratory usually also has ISO 9001 certification.

This is due to the uniformly oriented specifications for process-oriented implementation. However, these are generally valid in ISO 9001 and apply to both service-providing and manufacturing companies / enterprises.

Therefore, especially for calibration laboratories, ISO 10012, "Measurement management systems . Requirements for measurement processes and measuring equipment" was created specifically for calibration laboratories.

This regulation supplements the above-mentioned ISO 17025 and ISO 9001 and provides calibration laboratories with tools and guidelines for implementing calibration with high quality standards and comparable quality.

Metrology basics

Profile of a test-and measuring device – added value

A measuring and testing device must (be able to) be trusted - trusted to believe that it is within its specifications and will always measure "accurately" when used properly and professionally.

In a technically oriented environment full of highly complex systems, trust - at least in the colloquial sense - is not a factor on which to build and plan.

Calibrations provide that confidence. Regular calibrations create a data-based and thus resilient foundation for such trust.

Calibration determines whether the measuring and testing instrument is within its specifications. If it does not, the instrument must be adjusted or repaired.

Data on these measures are available to the device owner as a result of this calibration - for example in the form of a calibration certificate.

Here, a value creation has already begun that can be sustainable: if the data of several - at least three calibrations - are compiled and evaluated, the calibration interval can be monitored and adjusted and the measuring instrument is classified with regard to its reliability.

Only through an initial and then subsequent calibrations does the product "measuring instrument" acquire its metrological profile, which allows statements to be made about its reliability and its precision.

In order to develop such a profile, it is not absolutely necessary, but strongly recommended, to always use the same calibration equipment. This ensures the same conditions and procedures during calibration. "Jumping" between different calibration facilities, e.g. for cost reasons, often does not allow an evaluation due to non-comparable calibration scopes or non-uniform calibration certificates.

Replacing a torque wrench for cost reasons

The following argument is often heard: *"...just throw the item away; a new purchase is cheaper than a calibration..."*.

These arguments are often used for cheaper measuring devices such as torque wrenches, multimeters, feeler gauges, dial gauges, etc. .
Those who argue in this way have not understood that they are talking about two products:
1. a measuring device (the "hardware") and
2. the determination and specification of the (metrological) properties of this device.

A measuring device is often a mass product - in many applications it cannot be exchanged at will - it is a product which has been manufactured and whose very specific properties - in the case of a measuring device a dimensional standard - are only documented by a precise initial and subsequent measurement (i.e. by a calibration).

Only if calibrations are carried out regularly by an (accredited) calibration laboratory, can knowledge be gained about
- the reliability
- the precision
- the stability
of the instrument, to build up a history and to make estimates about stability and calibration interval.

Torque wrenches - basics and calibration

Traceability

In connection with calibrations and measuring and test equipment, the term "traceability" is mentioned again and again and traceability is required.
Simply explained, traceability means that any measuring instrument is calibrated against a more accurate measuring instrument (a standard), which in turn has been calibrated against an even more accurate standard - this chain is to be continued until the most accurate standard available - usually a national standard, e.g. at US NIST or German Physikalisch-Technische Bundesanstalt - is reached.

In the process, a whole series of characteristics must be documented at each stage in order to prove that the national standard has been reached and thus to demonstrate traceability.

The formal definition of metrological traceability:
"Property of a measurement result whereby the result can be related to a reference through a documented unbroken chain of calibrations, each contributing to the measurement uncertainty."

To receive "traceability", an accredited calibration laboratory is highly recommended / necessary.

To ensure (audit) that the laboratory is traceable, this means for example, but not limited to, the following:

- traceability of that laboratory is documented
- its quality system and proper procedures are in working order
- competence of workers are adequate
- uncertainty of the calibration is properly calculated
- uncertainty of the calibration is suitable for your use

To find out all the necessary information, it requires a very good dedicated competence of the person performing the audit of the laboratory.
If the laboratory is accredited, competent auditors are auditing the laboratory on a regular basis, ensuring the 17025 requirements are met.

The ILAC P-10:2002 (refer to NIST homepage www.nist.gov) considers the elements of conforming metrological traceability to be
- an unbroken metrological traceability chain
- to an international measurement standard
- or a national measurement standard
- a documented measurement uncertainty
- a documented measurement procedure
- accredited technical competence
- metrological traceability to the SI
and calibration intervals

Torque wrenches - basics and calibration

Traceability:

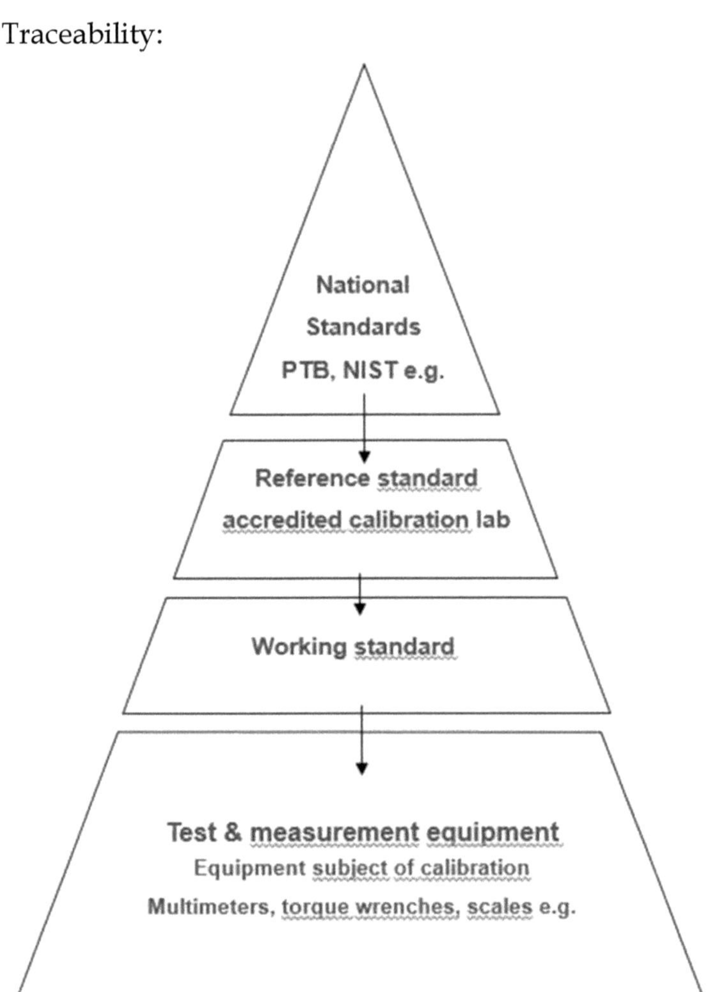

Documentation

International vocabulary of metrology (VIM): *"... where the result is supported by a documented, ...*
"

As with all elements of a QM system, written proof of all process steps is essential. This applies to the proof of
This applies to the proof of calibrations performed as well as to the proof that these calibrations were performed on a standard, which in turn was calibrated on a calibrated standard.
Accordingly, a measurement result has further properties in addition to a determined measured value. These properties also include - consistently carried out - a relationship to one or more further measurement results and respectively assigned measurement uncertainties. A determined / read measured value does not stand alone and absolutely in space, but belongs to a part of a chain or cascade of further measurements, measurement uncertainties and measurement results when considered correctly and comprehensively.

Calibration hierarchy

International vocabulary of metrology (VIM):
"... unbroken chain of calibrations,..., can be referred to a reference".

Test- and measurement equipment in your own plant or company might be tested or even calibrated on an own test- or calibration system. This system performs the function of a factory or working standard.
Of course, this system must also be calibrated at regular intervals. This chain can / must be continued.

The next higher stage is the connection via an accredited calibration laboratory. Typically, after about three or four stages, one arrives at the highest available standard - usually at a national standard - in the Federal Republic of Germany at the Physikalisch-Technische Bundesanstalt in Braunschwei or in the US at NIST.

This chain of relations from measuring instrument - standard to the national standard is called traceability chain in metrological parlance.

The dependencies of this traceability chain are clearly shown in the diagram on page 89. The direction of the arrow stands for the forwarding direction of the standard results.

The test- and measuring equipment as often found in larger quantities was designated as utility material; e.g. multimeters, micrometer gages, oscilloscopes, torque wrenches, etc.

A working or factory standard is often found in those companies, e.g. with a calibration system for torque wrenches.

To meet traceability requirements, these standards must be calibrated by an accredited calibration laboratory - they require a calibration certificate issued by an accredited laboratory.

Some large companies even have their own factory calibration laboratories. For the accredited parameters, these laboratories can perform the calibration of the factory standards.
The standards of this calibration laboratory are to be calibrated either directly by a national institute are or another competent accredited calibration laboratory.

Measurement uncertainty

According to the International vocabulary of metrology (VIM):

"... of calibrations, each of which contributes to the uncertainty of measurement...."

Traceability, according to the definition quoted above, is not defined exclusively by the measuring instrument, but by the measured values that are obtained.

Which components does a measurement result consist of?

- A measurement leads in almost every case to a numerical value (e.g. read value).
- This numerical value is assigned a unit (e.g.: Newtonmeter .).
- Unfortunately, there is always an error - no measurement is "infinitely accurate".

Expressed as a formula, this looks like this:

$$X_w = X \pm E$$

with

X_w as „true" measured value
X as measured value
E als error.

If one knew this error, it would be easy to convert the measured value into a true measurement result.

Unfortunately, the true value of this error is never known - by suitable containment of as many known or suspected error components as possible (examples: Precision / "measurement accuracy" of the measuring device, temperature influences, measurement conditions, measurement procedure, etc.), however, this error can be described and kept as small as possible.
Accordingly, a whole series of attributes can be assigned to a measured numerical value.

One attribute is the measurement uncertainty, in which the described error influences are summarized and which falsify the determined numerical value. In summary, the measured value and the included measurement uncertainty is a measurement result.

Measurement uncertainty - short introduction

"You actually measure wrong all the time, you just have to know how much." [Dave Packard]

As already shown in the "Traceability" chapter: errors and uncertainties are made with each measurement.

Conclusion:
There is always an error in a measurement result - no measurement is "infinitely accurate".

Once again expressed as a formula:
$$X_w = X \pm F$$
with

      X_w as „true" measured value
      X as measured value
      E als error.

So, some measurement uncertainty is always part of a measurement result. It is the "error" inherent in a calibration:
- the standard, against which is compared, has an inaccuracy, too
- the measurement procedure may contain (small) errors
- errors / uncertainties of the evaluation software
- Error influences by environment (temperature, humidity or similar)...

Torque wrenches - basics and calibration

These "errors", better called deviations, can consist of the following components:

Systematic deviations

Systematic deviation are typically known quantities. They can usually be corrected. If this is not possible, these deviations are to be added linearly.

Random deviations

Random deviations must be taken into account on the basis of statistical calculation principles. Therefore, in this context, we also speak of an estimation.
These estimates are now assigned a confidence interval, which is typically 95%: this means that 95 of 100 measurements lie within this estimated range.

In order to raise the now determined measurement uncertainty, which still contains a considerable unreliability, to a safe level, a coverage factor is included. This factor is typically assumed to be k = 2.

This factor comes from Gaussian normal distribution theory and is exactly k = 1.96.

A simple example will illustrate this:

The idea in stating a measurement result is:

**The statement about the measurement result
MUST be correct = true.**

Example:
The width of a door is to be determined. A Class III wooden folding ruler is used to read off a value of 79.8 centimeters. As explained, this value is only part of the measurement result.

This value must now be assigned a measurement uncertainty. In the simplest case, this measurement uncertainty is determined by setting up a matrix of possible error influences. The error is then assigned to the individual positions in this list. If this error is not known, it may also be estimated (this worsens the numerical value of the result - i.e. the measurement uncertainty increases), but improves the overall result because it is truer.

Example: folding ruler:

Error matrix "Folding rule"	
Basic accuracy	0,6 mm
Accuracy class III	0,4 mm/m
Temperature / Environment	± 2 mm (geschätzt)
Hinge clearance	± 0,5 mm
Handling (reading error, "odd" application)	± 1 mm

First, the error limit of the link joint bar dimension must be determined:

$$a + b * L$$

with

- L = value of the length to be measured, rounded up to the next full meter
- a, b to be taken from the accuracy classes for linear encoders, EC Directive 2004/22/EC

Note: the error limit is not to be confused with the measurement uncertainty!

In the considered example this results in:
0.6 mm + 0.4 mm/m * 1 m = 1.0 mm

Now the measurement uncertainty can be calculated: The individual errors are now squared and added.

The root of this sum is the determined measurement uncertainty (geometric addition).

When adding measurement uncertainties, ensure that all components have the same dimensions!

$$u = \sqrt{1^2 + 2^2 + 0.5^2 + 1^2 + 1^2} = 2.693 \text{ mm}$$

Now, the extension factor will be included:

$$2.69 \text{ mm} * 1.96 = 5.27 \text{ mm}$$

The "exact" measurement of 79.8 cm therefore has a measurement uncertainty of 5.27 mm.
This seems too high? If one considers the individual components and then takes into account an interaction under the worst possible circumstances: inaccurate application, inaccurate reading, play in the hinges, extreme e.g. summer temperatures, etc., the value is not so unrealistic.

However, there is suitable software for setting up model equations and calculating the expanded measurement uncertainty.

The GUM is considered as a reference:
GUM is the abbreviation for the ISO/BIPM guide "Guide to the Expression of Uncertainty in Measurement".

It was first published in 1993 and last revised in 2008. The authoritative German version is the pre-standard DIN V ENV 13005 (current edition:1999-06) "Guide to the Expression of Uncertainty in Measurement".

The GUM was implemented in a software "GUM Workbench". More detailed information is available at http://www.metrodata.de/.

A free online calculation of measurement uncertainties can be done e.g. via the homepage of NIST (National Institute Of Standards), http://uncertainty.nist.gov/ .

Uncertainty of measurement or tolerance?

Unfortunately, there is always confusion around these two terms. If a measuring device has a tolerance (example: ± 3%), what is the measurement uncertainty?

When a quality product is manufactured, it is guaranteed to have specified properties. An example - deliberately not chosen for a measuring instrument - shall clarify:

A door with a width of 90 cm is to be manufactured. A (fictitious) specification requires a manufacturing tolerance of ± 1%.

This means that if this specification is met, no door will leave the factory that does not have a width between 89.10 and 90.90 cm:

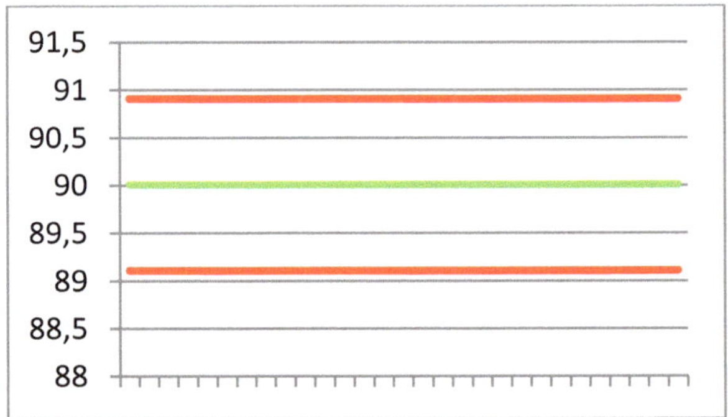

X-axis: Number n of doors manufactured
Y-axis: Width in cm

As long as the dimensions of the door are between the two red lines, the door is within the agreed / defined specifications.

This now has to be proven - i.e. re-measured for each door as part of production. A suitable measuring device is used for this purpose.

This measuring device also has a basic tolerance. It is clever to choose a measuring device that is clearly "better (m)i(s)st" than the permitted tolerance of the door.

As described in the chapter "Measurement Uncertainty", a measurement result is composed of a number of components - the tolerance of the measuring device is only one factor.

Assuming an expanded measurement uncertainty of ± 0.1 cm is reported for the measuring device - i.e. almost 10 x better than the permissible tolerance of the door - then safe production should be ensured:

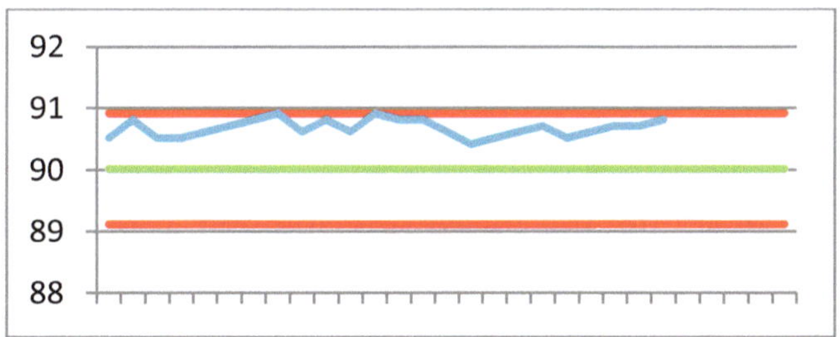

All doors produced are safely "in tolerance" - some of them do exhaust the
Although some of them exceed the maximum tolerance range, they are within the permissible limits,

But if you now show error bars for the measurement uncertainty of the measuring device used, you can see:

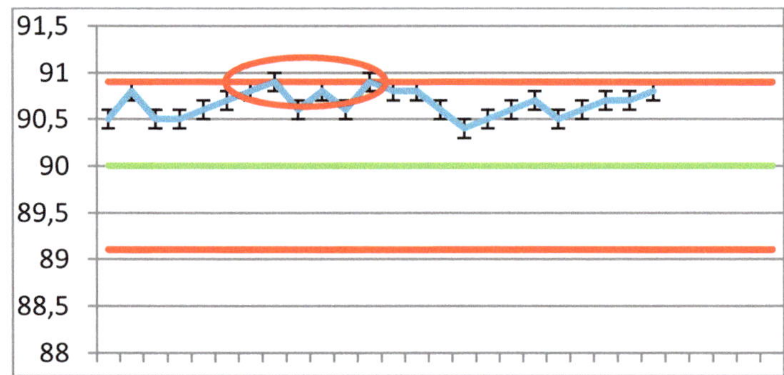

With the doors at the upper edge of the tolerance, the statements "in tolerance" cannot be made with any certainty at all! Because the measurement result has an uncertainty, it could well be that some of the doors produced are out of tolerance!

In this example, the measuring method / measuring device or the specifications must be improved!

Torque wrenches - basics and calibration

For the evaluation of the suitability of a calibration applies:

- The product should not better than the standard
- Plain sight on the product is not good enough for a qualitative assessment
- Either the scale is adjusted - or a calibration with smaller measurement uncertainty must be demanded!

Another view on tolerances and measurement uncertainty shall be demonstrated by means of the already known targets:

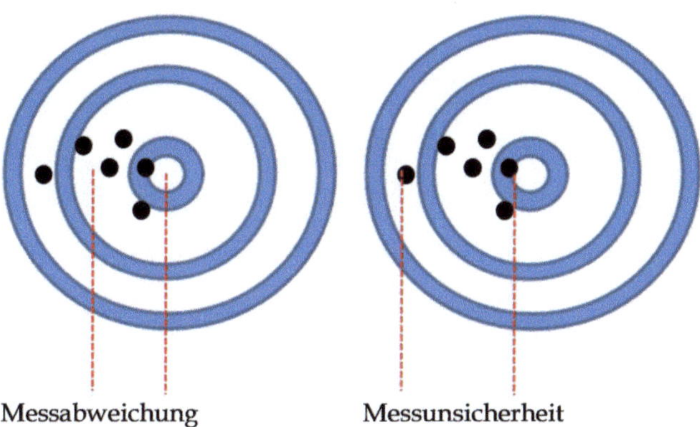

- The specification of a tolerance describes the (assured or desired) quality of a product.

- The measurement uncertainty describes the quality of the measurement.

When looking for / selecting a calibration service, it is therefore necessary to assess whether the quality of the calibration - expressed, among other things, in terms of measurement uncertainty (other characteristics are, for example, number of measurement points, direction, etc.) is sufficient for one's own measuring device.

Labeling of test- and measuring equipment

ISO 9001:2015 requires in section 7.1.5.2 ("Metrological traceability", excerpt):

"Where metrological traceability is a requirement, or is considered by the organization to be an essential contribution to establishing confidence in the validity of measurement results, the measuring equipment shall:

> *(a) be calibrated, verified, or both, against standards traceable to international or national standards at specified intervals or prior to use; if such standards do not exist, the basis for calibration or verification shall be retained as documented information;*

> *(b) Be marked to allow their status to be determined;*

> *c) ... "*

Translated with www.DeepL.com/Translator (free version)
The consistent implementation of this (considered very reasonable) requirement means:
Mark each measuring or test device of your company with this sticker, which must be labeled / filled in according to the entries of the measuring equipment monitoring.

Recommendation: Design a calibration mark (also called calibration sticker or calibration sticker) for your operation.

Such a uniform label has some decisive advantages:

- each employee can be instructed to always check the calibration sticker for validity before starting work

- with a large number of measuring and testing devices, it is not unusual for measuring devices to be temporarily untraceable or inaccessible. This can happen due to internal relocations, rental of the devices, infrequent use, when issued to field staff, etc. If such a measuring device reappears, the calibration status can be determined by a quick glance at the calibration sticker.

- If measuring and test equipment is assigned to external calibration services for calibration, you will usually receive a calibration mark from the laboratory performing the calibration. This can be a label of an accredited calibration laboratory as well as - in the case of service calibrations - be any mark of the calibration facility. The following applies to all these calibration stickers:
Nothing is uniform. Different formats, colors, and information mean that the contents can only be poorly understood and implemented by the user. If a company's own sticker is applied in addition to these "foreign" stickers, this weak point is compensated for and the above-mentioned first line enumeration can be demanded with emphasis.

A simple calibration sticker can also be obtained for little money. Even if there are no limits to your imagination, you should limit the number of fields to the most necessary - after all, the sticker should be able to be detected "at a glance".

Torque wrenches - basics and calibration

Example of simple sticker :

Technician
OrderNo:
Calibrated on / due

This sticker, which is only approx. 2 x 1 inch in size, contains all the important details which
- establish references to the calibration laboratory ("calibrated by", "order number") and the last calibration.
- indicate the current calibration status

Recommendation for filling in such a calibration sticker:

Line 1:	(optional): Add your companies name/logo
line 2:	name of the calibration laboratory, department if applicable
Line 3:	Order number
Line 4:	Field "Next calibration":

By specifying the calibration center and the order number, a new calibration can be requested if necessary.
- a new calibration can be requested if necessary
- the corresponding calibration certificate can be found.

Each calibration body has an order or operation management system; an accredited calibration body is even obliged to keep calibration results (compare: ISO 17025:2018, item 7.5 "Technical records").

With the respective order number, the last order can be called up - all device data (type, serial number, etc.) are thus available to the calibration laboratory, and a new calibration can be prepared or offered. When "ordering" a calibration, e.g. for a multimeter, it may be the calibration for one of 10 multimeters of the company - the calibration center may calibrate several dozens of multimeters of this type per day - here the exact address of the multimeter is helpful to obtain the desired calibration scope and quality.

Torque wrenches - basics and calibration

In case of repeated calibrations of the measuring instrument, the title pages of the calibration certificate hardly show any changes - holder data, type and serial number usually remain unchanged. The date of creation and the order number change. If you have a central or decentralized repository for calibration certificates, the last calibration certificate must be easily identifiable - the order number is helpful here.

If the calibration certificate is not (or no longer) available, a copy can be requested from the calibration center, stating the order number.

Special attention must be paid to the 3rd line "Next calibration":
A calibration body may not indicate the time of the <u>next</u> calibration:

ISO 17025:2018, Clause 7.8.4.3 "Special requirements for calibration certificates":
A calibration certificate or a calibration mark shall not contain a recommendation on the calibration interval], <u>unless this is done with the consent</u> of the customer.

The aim here is to prevent unwanted "customer loyalty": it is basically the free decision of the instrument holder whether, when and how often he presents his instrument for calibration.

This free decision may be restricted by requirements such as those contained in the Product Liability Act or ISO9001 - nevertheless, the equipment owner should remain free in his decision and not be bound to a calibration facility.

On the other hand, in view of these standards and laws, there is of course also the obligation to fall short of a calibration interval if, due to failure, malfunction or breakage, the validity of the last calibration result must be doubted.

Having your own calibration sticker (which, can be provided to a calibration laboratory for use on your equipment!) has several advantages:
- The agreement of the indication of a calibration interval is made by providing it,
- as explained above, each employee can be instructed to check the validity of the (company-owned) calibration stickers in principle before starting his work

Unfortunately, there is no guideline (yet) for a uniform or standardized calibration sticker. The information required above can only be read with difficulty or often not at all from many calibration stickers used.

Inconsistent designs, poor legibility, seemingly arbitrary information found to be necessary do not help and do not support modern gauge management and do not meet the standard's requirement for labeling.

With this variety, it is impossible to expect that, as described above, the calibration status of the measuring device can be detected "at a glance". An auditor will also have a hard time and immediately demand calibration certificates.

Basically, it will not be possible to dictate to any calibration service which sticker is desired. The (company's) own calibration sticker, which is affixed instead of or in addition to the calibration sticker of the calibration body, is a remedy. This is particularly recommended for stickers of accredited laboratories; these should not be removed under any circumstances.

Torque wrenches - basics and calibration

Calibration label (German accreditation body DAkkS e.g.):

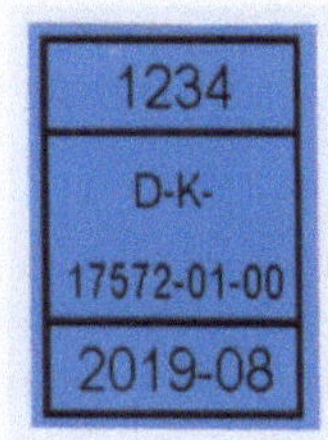

Size:	18 x 28 mm
Color:	blue
Line 1:	Count number / consecutive calibration number.
Line 2:	Registration number of the accredited calibration laboratory
Line 3:	Year and month of calibration (date of performance).

With regard to further standardization, however, there are further efforts: here it is the military, which has recognized through joint operations that standardization is indispensable in order to overcome language, structure and even writing barriers.

With the NATO paper STANAG Nr 4704 "NATO Requirements for Calibration Support of Test and Measurement Equipment", (1st edition) a framework for a standardized sticker was defined.

As with many quality-related specifications from the military sector, it can be assumed that in a few years' time a corresponding standard or normative requirement is also implemented for the civilian sector.

The calibration sticker shown as an example in this chapter already includes parts of the STANAG requirements.

Calibration stickers should be attached to the front or top of the device so that they are clearly visible. For small gauges or gauges without space for a sticker, it can also be attached to the associated case / stowage container (example: gauge block set: not every gauge block receives a sticker, the entire set carries a sticker on the usual wooden storage case).

Calibration intervals

Indication of the recalibration time

A question that is often discussed is: when exactly does the validity of the calibration expire? (Compare also above mentioned remarks on "Interval specifications in the calibration certificate").
There is no official or standard-based specification to answer this question.
If a measuring device has a calibration interval of 12 months and the last calibration was on 16.04.2015 - when does it have to be calibrated again?
There are operational management systems that are also used for monitoring measuring equipment that keep a complete date.

According to this, the measuring device should no longer be used from 16.04.2016 if it has not been recalibrated beforehand.

It would be better - because it is more practical - to make a stipulation that a measuring device must always be presented for calibration on a monthly basis. In our example, this would be in April 2016.
In this way, a maximum of almost a full month can be "gained". This only appears at first glance as additional time of use. As a rule, the re-presentation for calibration cannot be coordinated to the exact day - here there are to many imponderables, which can lie

in the own company, but can also lie with an external calibration service provider.

Choosing the "target month" as the expiry date therefore merely gives the necessary leeway for everyday registration for calibration.

If one decides on this procedure
- it must be written down in the quality management manual
- the re-presentation date in the measuring equipment monitoring system must be kept accordingly
- the own calibration sticker (if used) must be filled accordingly only with indication of month and year.

Start of calibration interval

The expiration of a calibration interval starts counting from the date of calibration. It is a common misconception that the time only runs from the first use.
It is certainly annoying that part of the usage period must be set aside for logistical operations such as shipping. But it must be understood that it is not usage that affects a meter, but many factors. These include influences that can have negative effects on the gauge during storage or "no use".
Examples are already listed in the section "No interval extension".

Pausing use

Pausing the use of a measuring instrument and, for example, storing it for a period of time does not extend the calibration interval.
calibration interval. The interval begins with the time of completion of a calibration.

Calibration interval of torque wrenches

ISO 6789-1:2017 specifies interval requirements:

If the user utilizes procedures for the control of test devices, torque tools shall be included in these procedures. The interval between conformance tests shall be chosen on the basis of the factors of operation such as required maximum permissible deviation, frequency of use, typical load during operation as well as ambient conditions during operation and storage conditions. The interval shall be adapted according to the procedures specified for the control of test devices and by evaluating the results gained during successive conformance tests.

If the user does not utilize a control procedure, a period of 12 months, or 5 000 cycles, whichever occurs first, may be taken as default values for the interval between conformance tests. The interval starts with the first use of the torque tool.

A shorter interval between conformance tests may be used if required by the user, their customer or by legislation.

With a 12 month /or 5000 cycle requirement for recalibration; a clear statement is made.

Torque wrenches - basics and calibration

Questionable is the statement *"The interval starts with the first use of the torque tool.2*
It has been experienced, that torque wrenches, which have been stored for a longer period of time, malfunctioned caused by resinification / hardening of the internal lubricants.

General information on calibrations and calibration intervals

Approximately 8% of all measuring and test equipment must be set and adjusted during a calibration.

Torque wrenches in general are often used in rough conditions. Even they might appear as about unbreakable, many wrenches are sensitive instruments and need to be readjusted to meet specifications.

However, a clear demarcation must be drawn:
If an instrument is presented as "defective" to the manufacturer or a calibration facility (often identical), these items were not taken into account.

The owner of the test&measurement device owner knew already that there was a failure (failure, malfunction or breakage).

Included were instruments that were "just" supposed to be calibrated and that the user did not believe were measuring incorrectly or inaccurately.

Although this figure has a certain fuzziness, it is based on an evaluation of a large pool of devices (> 200,000 devices, more than 10,000 different types of devices of all kinds: from simple multimeters to spectrum analyzers, from coarse scales to low-pressure standards); coincides with experiences of other device owners with a large amount of measuring and testing devices and can be used as a "thumb value" and thus as a guideline for an own target of a quality score.

Calibration scheduling

Calibration services requirements

The calibration facility cannot know the needs / wishes of the torque wrench holder.
A simple sending in of a wrench "for calibration" always results in inquiries, which altogether lead to delayed processing.

The calibration laboratory always needs the following information:

In which measurand is a calibration needed?
- Torque or also angle of rotation?

Type of calibration :
- Factory calibration or traceable calibration?

Required direction of rotation of the calibration:
- only direction of rotation right (cw)
- Direction of rotation right and left (cw + ccw)

Torque wrench type and calibration requirement:
- indicating without b-parameter determination
- indicating with b-parameter determination
- triggering without b-parameter determination
- triggering with b-parameter determination

Torque wrench data:
- Specification of type and serial number of the torque wrench.
- For calibrations that are NOT performed at the manufacturer: Submission of the technical data for a conformity statement (e.g. data sheet).

Especially the last point is important if a conformity statement is required.

A manufacturers calibration laboratory usually knows the characteristics of the torque wrench.

An independent calibration laboratory may not know this information. As explained in the chapter "Conformity statement", there are numerous possibilities to agree on a decision rule.

The reference to "manufacturer's data" does not help the calibration laboratory either: it must research and obtain the data. One can readily find different data sheets on measuring equipment on the Internet: A successful type may well have slightly different data sheets over the years - if you find the data sheet at the manufacturer, this may well differ from the data sheet of a tool wholesaler. The decision as to which data sheet should be the basis for a conformity statement must lie with the torque wrench holder.

An example of a checklist "Calibration torque wrench" is on the following page.

Torque wrenches - basics and calibration

Checklist calibration

Checklist „**Torque wrench calibration**"

Required measurand of the calibration:
- ☐ Torque ☐ Angle of rotation:

Calibration type :
- ☐ Service calibration
- ☐ traceable calibration

Required dirction of rotation:
- ☐ clockwise only (cw)
- ☐ clockwise and counterclockwise
  (cw + ccw)

Torque wrench type and calibration requirement:
- ☐ indicating w/o b-parameter determination
- ☐ indicating w b-parameter determination
- ☐ releasing w/o b-parameter determination
- ☐ releasing w b-parameter determination

Torque wrench data:
- Specification of type and serial number of the torque wrench.
- For calibrations that are NOT performed at the manufacturer: Submission of the technical data for which a statement of conformity is required (e.g. data sheet).)

Service- or traceable calibration

When planning / registering a torque wrench for calibration, one must face the question of whether a service calibration is sufficient or whether a traceable calibration is required.

A frequently encountered - but unreasonable approach - is often the question of costs. The answer can be given as a blanket statement: a service calibration is often significantly less expensive. This often leads to the decision in favor of this type of calibration.
Such an approach is too short-sighted and often becomes a problem during the next audit.

There is no official definition for the term service calibration, also known as factory or manufacturer or company calibration.

Thus, at first, one does not know exactly what is offered. A service calibration may be performed based on standards and regulations or on the basis of the customer's requirements – or on very own ideas of a calibration.

This may be sufficient for many applications, but the decision for a factory calibration should be thought through in a structured way and subsequently based on a decision matrix.

Germany's accreditation body DAkkS publication "DAkkS-DKD 4" defines factory/manufacturers calibration as follows:

"In-house calibration (factory calibration)
6.4.1 An in-house calibration system ensures that all measuring and test equipment used in a company is regularly calibrated with the company's own reference standards. For the company's reference standards, traceability of measurements shall be ensured by calibration in an accredited calibration laboratory or a metrological state institute. In-house calibration may be evidenced by an in-house calibration certificate, calibration mark or other suitable method. Calibration records shall be retained for a prescribed period of time.

6.4.2 The type and scope of metrological control in the case of in-house calibration are left to the company concerned. They shall be adapted to the particular applications so that the results obtained with the measuring and test equipment are are sufficiently accurate and reliable. Accreditation of organizations performing in-house calibrations is not required to meet the requirements of the EN ISO 9000 series of standards applied to in-house issues.

However, if an external body uses an in-house calibration certificate as evidence of traceability, it should be required that the issuing organization can demonstrate its competence."

An analysis of this text results for the implementation:

- A factory/manufacturers calibration certificate is regarded as an in-house calibration certificate.
- If the factory calibration certificate of another, external calibration body is to be recognized / used, this body should be accredited.

When selecting a suitable calibration laboratory for a factory calibration, a calibration body that is accredited in at least one parameter according to EN 17025 should be the first selection criterion:
A calibration laboratory that has such an accreditation has a quality management system and basically has to comply with the whole range of specifications of this standard.
It is possible, but extremely unlikely, that a laboratory that goes to the trouble of accreditation has a parallel system, which leaves all organizational and also infrastructural expenses lying around and carries out a "backyard calibration" without any quality-relevant requirement.

The second selection criterion should be the determination of what is measured with the measuring and testing instrument. In this case, it is not the measured quantities that are meant, but the application areas of the instruments.

The following questions should be answered:

- Is the measuring instrument a device for daily use (occasional measurements: e.g. caliper to determine the thickness of a drill bit, multimeter to determine the voltage lead of a socket, etc.).

- Or are regular, critical or quality-relevant measurements carried out (e.g. re-measurement of production masses)?

- Or is the measuring instrument even used for calibration (checking and calibrating other measuring instruments in the company, e.g. multimeters or torque wrenches)?

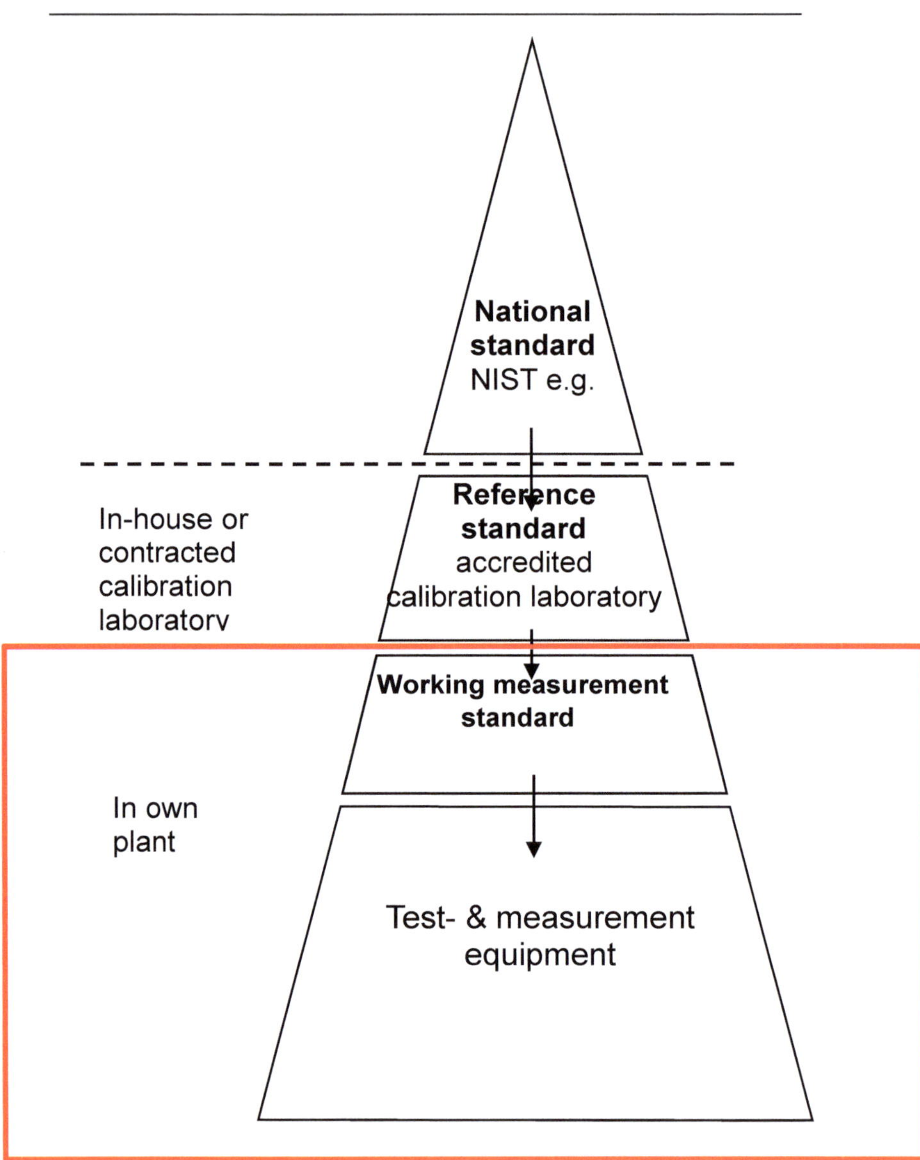

National
standard
NIST e.g.

Reference
standard
accredited
calibration laboratory

In-house or
contracted
calibration
laboratory

Working measurement
standard

In own
plant

Test- & measurement
equipment

To answer these questions, one should group the measuring and testing devices with regard to their field of application.
This grouping is most easily done using the pyramid already presented to visualize one's own calibration hierarchy.

Arrange (the measuring and test equipment in question) in the appropriate levels of the pyramid - as a rule, only the two lower levels come into question, level three can occur at individual companies.

If one examines the measuring and test devices of one's own company in this way, one may find that perhaps even working standards are available or that the purchase may be economical.

If a large number of torque wrenches requiring calibration are available, it is well worth considering procuring a (traceably calibrated calibration device) and performing the wrench calibrations as factory calibrations in one's own company.

DAkkS-DKD-4 provides precise information here:

"Working standards or factory standards must be calibrated by an accredited calibration laboratory and thus be traceable to national standards."

This provides a clear decision-making aid.

Torque wrenches - basics and calibration

Summary: The difference between "real" traceable calibrations and factory calibrations can be that applicable standards or regulations for calibration (e.g. VDI/VDE 2646 for torque sensors) are only applied in a simple, "slimmed down" form. This means that simple, cost-effective and technically correct calibrations can be carried out.

In the case of traceable calibrations, the entire calibration procedure, including the form and content of the calibration certificates, is prescribed by the accreditation body / (inter-)national norms and standards, which leads to a considerably higher effort and confidence for these calibrations.

For the creation of factory calibration certificates, there are not specifications for all parameters and then lies in the discretion of the calibration laboratory.
In the case of services calibration certificates, it is therefore essential to ensure that the measurement uncertainty as well as the proportions leading to the measurement uncertainty are stated.
Calibrations without indication of the measurement uncertainty are worthless and decided as comparisons or functional tests.
Whether factory or traceable calibrations are necessary is decided by a calibration hierarchy that should be created for the company.

Calibration results

Post- calibration checks

If the torque wrench is returned after calibration, it should be checked as a matter of urgency. In a calibration laboratory, metrological but also, for example, formal errors can occur - here, too, "only" people are at work - it is advisable to create a checklist:

- Was a calibration certificate supplied?
- Is the information in the calibration certificate correct?
- Instrument data such as serial number
- order number
- Customer/client: if a service provider has been commissioned to monitor the measuring equipment and has also commissioned the calibration laboratory, the name of the service provider must not appear on the calibration certificate: the name of the equipment owner must be listed.
- Does the calibration scope documented in the calibration certificate match the commissioning?
- Is the calibration certificate metrologically plausible? Are the measuring range and/or measuring directions correct?
- Are recalibration data specified in the listing of the standards used and are they still valid?

Checks on the device:
- Has a calibration mark been applied?
- Are the entries on the mark correct?
- Are the returned accessories (ratchet, adapter, etc.) complete?

The most important step, however, is the evaluation of the calibration certificate:

It must be checked
- whether the torque wrench is / has been within specifications
- if necessary, an adjustment had been made.

If it becomes apparent that the torque wrench was / is out of tolerance, measures must be initiated immediately. This can be the recall of products that were screwed with the affected wrench or also a rework of workpieces.

Good and responsible calibration facilities will give notice to the equipment owner if irregularities are found.

In the event that a torque wrench is found to be "out of specifications", it is essential that the quality management manual contains a chapter on the measures to be taken.

A delimitation should be listed for the sake of completeness: The above only applies when a wrench is given for calibration in good faith that it is OK.

If a key fails due to failure, obvious malfunction or breakage, appropriate decisions can be made immediately - a calibration certificate after repair should not be waited for.

Calibration certificate

The scope of every calibration includes a document reporting des results - the calibration certificate.

The data that must be included in a calibration certificate can be found in ISO 17025:2017

According to ISO 17025, a calibration certificate must always contain:

- *a title (e.g. "Test Report", "Calibration Certificate" or "Report of Sampling") ;*

- *the name and address of the laboratory;*

- *the location of performance of the laboratory activities, including when performed at a customer facility or at sites away from the laboratory's permanent facilities, or in associated temporary or mobile facilities;*

- *unique identification that all its components are recognized as a portion of a complete report and a clear identification of the end;*

- *the name and contact information of the customer;*

- *identification of the method used;*

- *a description, unambiguous identification, and, when necessary, the condition of the item;*

- *the date of receipt of the test or calibration item[s), and the date of sampling, where this is critical to the validity and application of the results;*

- *the date(s) of performance of the laboratory activity;*

- *the date of issue of the report;*

- *reference to the sampling plan and sampling method used by the laboratory or other bodies where these are relevant to the validity or application of the results;*

- *a statement to the effect that the results relate only to the items tested, calibrated or sampled;*

- *the results with, where appropriate, the units of measurement;*

- *additions to, deviations, or exclusions from the method;*

- *identification of the person(s) authorizing the report;*

- *Clear identification when results are from external providers..*

Kalibrierlaboratorium für die Messgröße Drehmoment und Drehwinkel
Calibration laboratory for the measuring value torque and rotational angle

akkreditiert durch die / *accredited by the*

Deutsche Akkreditierungsstelle GmbH

als Kalibrierlaboratorium im / *as calibration laboratory in the*

Deutschen Kalibrierdienst

| 2345 |
| D-K-17572-01-00 |
| 2020-03 |

Kalibrierschein

Calibration certificate

Kalibrierzeichen
Calibration label

Gegenstand: *Object*	**Auslösender Drehmomentschlüssel** **Typ II Klasse A**	
Hersteller: *Manufacturer*	**Stahlwille** **Wuppertal**	
Typ: *Type*	**Manoskop 730D/20**	
Serien-Nr.: *Serial number*	**812460160**	
Auftraggeber: *Customer*	**Mustermann Produktion GmbH** **Musterstraße 1 - 4** **123456 Musterstadt**	
Auftragsnummer: *Order No.*	**123123987**	
Anzahl der Seiten des Kalibrierscheines: *Number of pages of the certificate*	**2**	
Datum der Kalibrierung: *Date of calibration*	**2020-03-15**	

Dieser Kalibrierschein dokumentiert die Rückführung auf nationale Normale zur Darstellung der Einheiten in Übereinstimmung mit dem internationalen Einheitensystem (SI). Die DAkkS ist Unterzeichner der multilateralen Übereinkommen der European co-operation for Accreditation (EA) und der International Laboratory Accreditation Cooperation (ILAC) zur gegenseitigen Anerkennung der Kalibrierscheine. Für die Einhaltung einer angemessenen Frist zur Wiederholung der Kalibrierung ist der Benutzer verantwortlich.

This calibration certificate documents the tractability to national standards, which realize the units of measurement according to the International System of Units (SI). The DAkkS is signatory to the multilateral agreements of the European co-operation for Accreditation (EA) and of the International Laboratory Accreditation Cooperation (ILAC) for the mutual recognition of calibration certificates. The user is obliged to have the object recalibrated at appropriate intervals.

Dieser Kalibrierschein darf nur vollständig und unverändert weiterverbreitet werden. Auszüge oder Änderungen bedürfen der Genehmigung sowohl der Deutschen Akkreditierungsstelle GmbH als auch des ausstellenden Kalibrierlaboratoriums. Kalibrierscheine ohne Unterschrift haben keine Gültigkeit.
This calibration certificate may not be reproduced other than in full except with the permission of both the Deutsche Akkreditierungsstelle GmbH and the issuing laboratory. Calibration certificates without signature are not valid. This calibration certificate is based on the german language. In case of doubt only the german version is valid.

Datum *Date*	Stellv. Leiter des Kalibrierlaboratoriums *Vice head of the calibration laboratory*	Bearbeiter *Person in charge*
2020-03-17		
	Michael Stader	F. Mauksch

Postanschrift/Mail address	Telefon-Durchwahl / Telephon-extension
Echantillon AG **Kalibrierlaboratorium** Fritz Maum Str. 71 D-42897 Remscheid	(+49) 02191 60199-0 0

In addition to these basic specifications, ISO 17025 specifies further requirements and devotes a separate chapter to these requirements in section 7.8.4:

7.8.4.1 In addition to the requirements listed in 7.8.2, calibration certificates shall include the following:

- *the measurement uncertainty of the measurement result presented in the same unit as that of the measurand or in a term relative to the measurand [e.g. percent);*

- *the conditions (e.g. environmental) under which the calibrations were made that have an influence on the measurement results;*

- *a statement identifying how the measurements are metrologically traceable [see Annex A);*

- *the results before and after any adjustment or repair, if available;*

- *where relevant, a statement of conformity with requirements or specifications (see 7.8.6);*

- *where appropriate, opinions and interpretations (see 7.8.7) .*

<table>
<tr><td colspan="3">2345</td></tr>
<tr><td colspan="3">D-K-17572-01-00</td></tr>
<tr><td colspan="3">2020-03</td></tr>
</table>

Seite 2 zum Kalibrierschein vom 2020-03-15

Page 2 of the calibration certificate of 2020-03-15

1 **Kalibriereinrichtung:** 1500-N-m-Dm-BNME # SCH-02

2 **Kalibrieranordnung:** Hebelarmlänge: 515 mm

Messachse: vertikal / vertical

Drehmomenteinleitung: Umschaltknarre VK 1/2" aus Lieferumfang, Stichmass 25,0mm

3.1 **Kalibriertemperatur:** 22,1 °C

3.2 **Luftfeuchte, relative:** 46,2 %

4 **Kalibrierverfahren nach DIN EN ISO 6789-2:2017 "Drehmomentschraubwerkzeuge Typ II Klasse A"**

5 Kalibrierergebnis Rechtsdrehmoment *Calibration results clockwise torque*

M_K N·m	$\overline{X}(M_K)$ N·m	relative Abw. $\overline{X}(M_K)$ %	$W_{KE}(M_K)$ %	$W"_{KG}(M_K)$ %
20,0	20,378	-1,855	0,20	2,908
120,0	121,958	-1,605	0,20	2,010
200,0	201,934	-0,958	0,20	1,238

M_K = Zielwert der Kalibriereinrichtung

$\overline{X}(M_K)$ = Mittelwert der Ablesungen

$W_{KE}(M_K)$ = relative erweiterte Messunsicherheit der Kalibriereinrichtung

$W"_{KG}(M_K)$ = relatives erweitertes Messunsicherheits-intervall des Kalibriergegenstandes

Angegeben ist die erweiterte Messunsicherheit, die sich aus der Standardmessunsicherheit durch Multiplikation mit dem Erweiterungsfaktor k = 2 ergibt. Der Wert der Messgröße liegt mit einer Wahrscheinlichkeit von 95 % im zugeordneten Werteintervall.

5.1 Bemerkungen / *remarks*

Das Prüfergebnis liegt im Rahmen der Aussage der Norm DIN EN ISO 6789-2:2017 innerhalb ± 4%, rückführbar auf nationale Normale.

6 Einzeldaten Rechtsdrehmoment *individual data clockwise torque*

M_K N·m	$X(M_K)$ N·m					
20,0		20,34	20,39	20,41	20,34	20,41
120,0		122,17	121,38	122,35	121,81	122,08
200,0	200,00	201,77	202,37	201,65	201,69	202,19
	Pre-load	1. Meas	2. Meas	3. Meas	4. Meas	5. Meas

7 Darstellung im Diagramm / *presentation in the diagram*

Interval specification in a calibration certificate

A recurring requirement of measuring instrument holders is the specification of a calibration interval or the specification of the time of the next calibration in a calibration certificate.

ISO/IEC 17025:2018 specifies in this regard:

7.8.4.3 A calibration certificate or a calibration mark shall not contain a recommendation about the calibration interval, unless this is done with the agreement of the customer.

The measuring device holder (respectively the QM system of the measuring device holder) alone knows how the measuring device is used (one-shift operation or "around the clock", laboratory conditions or construction site use) and must take these influences into account when assigning the calibration interval.
The specification of a calibration interval in a calibration certificate without the consent of the customer would mean an inadmissible interference of the calibration laboratory in the QM system of the measuring instrument holder. This could even lead to the assumption of an attempt to bind the customer - this would not be permissible.

Therefore, the interval specification must come from the measuring equipment holder; with this specification - if it is made known to the calibration laboratory preferably in written form - the approval is considered to be given.

Calibration marks for traceable calibrations / issued by accredited laboratories never have an interval specification or the specification of the next calibration.

It is permissible to affix a (own) factory sticker with interval indication or, better, the date of calibration / the date of the next calibration to the measuring equipment in addition to the traceable calibration sticker.

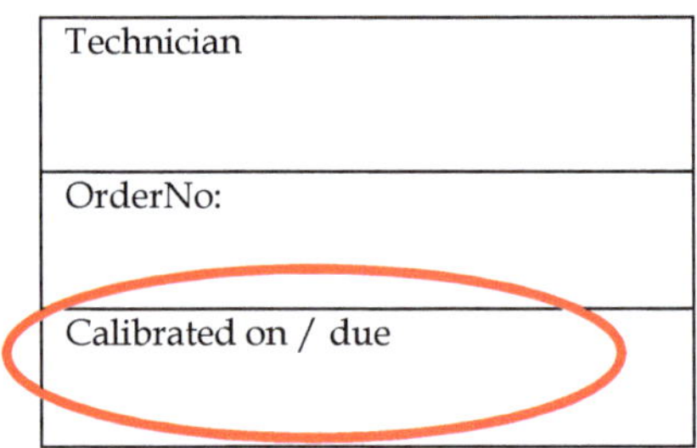

Torque wrenches - basics and calibration

ISO 6789:2 adds:

"If the user utilizes procedures for the control of test devices, torque tools shall be included in these procedures. The interval between calibrations shall be chosen on the basis of the factors of operation such as required maximum permissible measurement error, frequency of use, typical load during operation as well as ambient conditions during operation and storage conditions. The interval shall be adapted according to the procedures specified for the control of test devices and by evaluating the results gained during successive calibrations.

If the user does not utilize a control procedure, a period of 12 months, or 5 000 cycles, whichever occurs first, may be taken as default values for the interval between calibrations. The interval starts with the first use of the torque tool."

This last statement must be questioned emphatically. It cannot be a serious recommendation or specification to have a calibration interval start with the first use. There is evidence of numerous technical problems with new torque wrenches that were calibrated after production and then initially stored. This storage process had negative influences on the torque wrenches in large numbers: the lubricants used hardened and led to malfunctions.

To complete a further execution of the standard in this chapter:

"Shorter interval between calibrations may be used if required by the user, their customer or by legislation."

Statement of conformity

New in the 2017 edition of ISO 17025:2017 is the statement / specification about a conformity statement.

The inclusion of a conformity statement in the calibration certificate is based on the request of many measuring instrument holders and indicates whether a measuring instrument complies with specifications (e.g. the manufacturer's specifications) at the end of a calibration - or not.

In order to make such a decision, it must be determined what the "rule" is for this: the standard speaks of a decision rule.

There is no binding prescriptive rule for a conformity statement. This can be agreed between the calibration body and the instrument holder. Therefore, there are different models that can be agreed upon:

Normative / to be agreed specifications for the conformity statement:
- Conformity statement according to 14253-1
- Conformity statement according to ILAC G9 8-2009
- Conformity statement according to DAkkS-DKD-5
- Conformity statement without consideration of measurement uncertainty
- Conformity statement according to individual customer requirements

From this, conformity statements / the decision rule can be derived according to customer requirements, e.g.:
- without consideration of the measurement uncertainty
- "shared risk
- individual requirements

An agreement must be made with the calibration body if a conformity statement in the calibration certificate is desired.

The form of this agreement is currently practiced in different ways:
- The calibration laboratory has a note - possibly also as a footnote - in the order.
- The calibration laboratory has a flyer / information letter in which the decision rule is described, unless otherwise agreed.
- The general terms and conditions of the calibration laboratory contain a formulation.

If a statement of conformity is required by the calibration body in the calibration certificate, a specification of the decision rule should also be made - otherwise the communication in one of the forms described above or similar forms is regarded as an agreement!

The most common models of a decision rule are presented and explained below.

Decision rule – what's behind ?

The next page diagram shows possible cases of a measured value recording. The respective measured value lies on the dashed line, the double T stands for the measurement uncertainty of the measurement.

At the first measuring point, the measured value including the assigned measurement uncertainty is clearly within the specification limit.

In the second case, it can be seen that the measured value is clearly within the specification limit. However, due to the measurement uncertainty included, the value could possibly also be outside the specification limit.

In the third case, it can be seen that the measured value is clearly outside the specification limit. However, due to the measurement uncertainty included, the value could possibly also be within the specification limit.

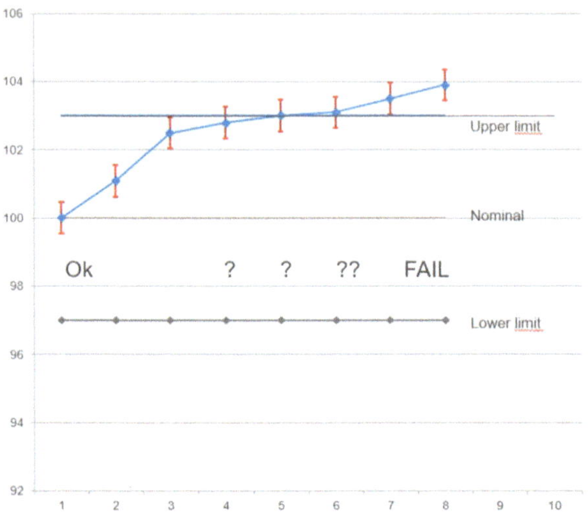

O.K.:   The instrument complies with the specifications, taking the measurement uncertainty into account.

?:   The measurement is within specifications / limits. Taking into account the measurement uncertainty, no statement can be made about compliance with the specification. The probability of compliance is greater than non-compliance.

??:   The measurement is outside the allowed limits. Taking into account the measurement uncertainty, no statement can be made about compliance with the specification. The probability of compliance is smaller than non-compliance.

Fail: The measurement result including the measurement uncertainty is greater than the upper error limit. The instrument does not comply with the specifications.

The measurement uncertainty, as described above, expresses the "quality" of the measurement or calibration.

Even if the measured value is "okay", i.e. within the specification limit, the actual result may be outside due to the uncertainty factor. The above graph shows that there are only two unambiguous results: The first measurement point is within specification including measurement uncertainty - the last measurement point is clearly outside specification. For the other two measuring points this cannot be determined with certainty.

ILAC G8 (ILAC: International Laboratory Accreditation Cooperation) implements these values as "cases":

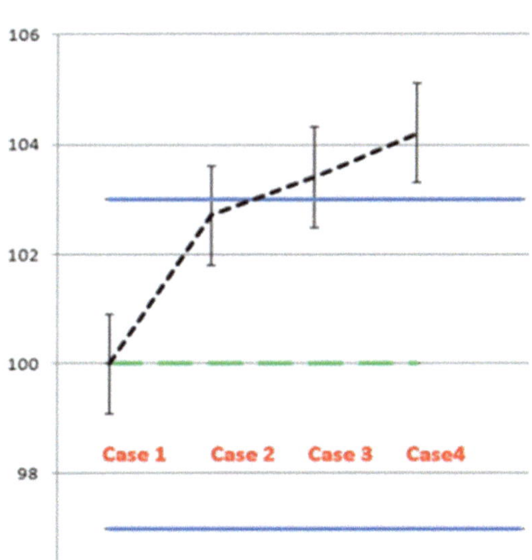

A clear statement on conformity or non-conformity can only be made for cases 1 and 4.

Must a torque wrench be compliant?

The purpose of a calibration is to detect a measurement deviation. Ideally, a measuring device would always display exactly the value that it actually measures.

In practice, this is never the case. But when is the instrument "in order" - when is it not?

All measuring instruments have three basic characteristic features:

- Precision
- Repeatability
- Linearity

If a measuring instrument always measures "wrong" at the same point of its characteristic curve, this is not a problem: with a correction table - in modern instruments by storage in a memory module with automatic correction of the display - the correct value can be determined.

Calibration certificates indicate the values determined over the measuring range - a correction of the measurement result can be done by the user.

This is too time-consuming or unpracticable for many users - there were requests to the standards committee to create a simple way of evaluating the calibration result. People want to be able to quickly see whether the instrument is as precise as they once bought it - the conformity statement was introduced.

However, there is still nothing to prevent working with the measured values determined during calibration - the device is not defective.
On older wrenches – produced before ISO 6789:2017 has been released, a special situation may occur:
While the previous 2003 version of ISO 6789 requested:

6.4 Calibration sequence
*The torque wrenches must first be **tested at 20%**, then at approximately 60%, and at 100% of the maximum torque value of the torque wrenches (or at the nominal value/adjustment value for torque wrenches*

the requirements have been changed to:

6.5.2 Indicating torque tools, Type I
Indicating torque tools (Type I, all Classes) shall first be measured at the lowest specified torque value of the measurement range (see 5.1.3) then at approximately 60 % and finally at 100 % of the torque tool's maximum value.

In some single case4s it may occur, that the engraved or printed nominal range starts with a lower value than the 20% value. This may lead to a non conformity statement. At this point, it is the instruments owner

either to accept that the lowest value is out of specifications – or redefine the specifications by defining own requirements for the decision rule.
Another easy way for a work around in that case is to order a calibration from the 20% value on (which is an own definition of the decision rule).

Note: there's nothing wrong with that wrench – it has never met the requirements in its life cycle – the new issue of ISO 6789:2017 now shows some unlinearity in the low range.

Room for big mistakes

A bad torque wrench or the wrong handling of a good torque wrench can have nasty consequences.

- Excessive torques overload the material, so that bolts either crack immediately or wear out more quickly due to material fatigue.

- Often, the threads are damaged and, in the worst case, the bolt gives way while driving.

- Even if the bolt holds, excessive torque can make it impossible to loosen it. Effort and costs incurred usually increase rapidly in such cases.

- Too little torque is also a serious source of danger. If, for example, the wheel bolts come loose because they have not been tightened properly, this can be life-threatening.

Insert torque wrench completely!

When things have to be done quickly: a torque wrench is not inserted all the way into the slip-on part / or the nut:

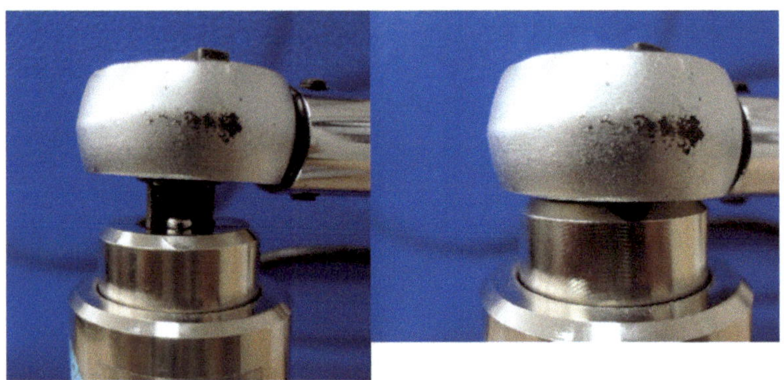

Consequence:

In addition to the risk of slipping and the associated risk of serious injury, the mount can burst - always at the weakest point of the square: at the corners - definitely not a warranty case!

Torque wrenches - basics and calibration

Insertion parts / accessories!

Don't save on accessories. A brand new torque wrench
was turned in for calibration – and failed!

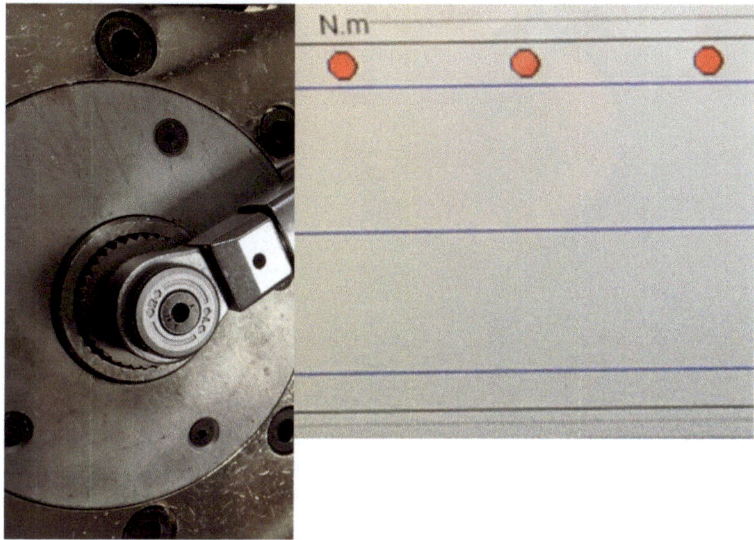

A closer (expert-)look :

The torque wrench owner saved some money by not
buying the matching accessory parts but using an
insertion part he had already

Unfortunately this part had a slightly different length
and hall length of the torque wrench.

Picture shows an original part and the presented part:

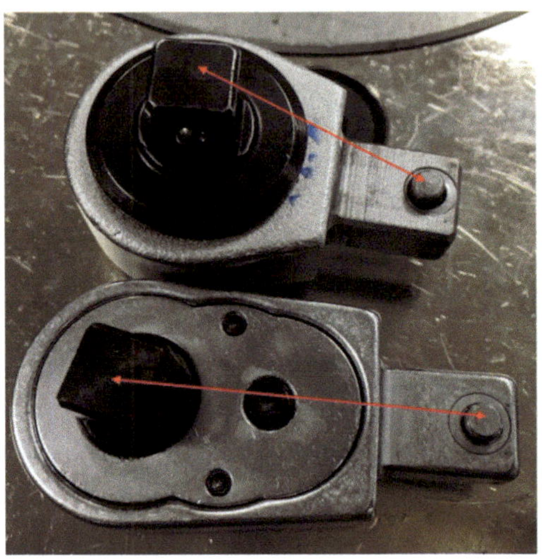

Wrong length of the tool head changes the overall length and the torque applied on the job is – wrong!

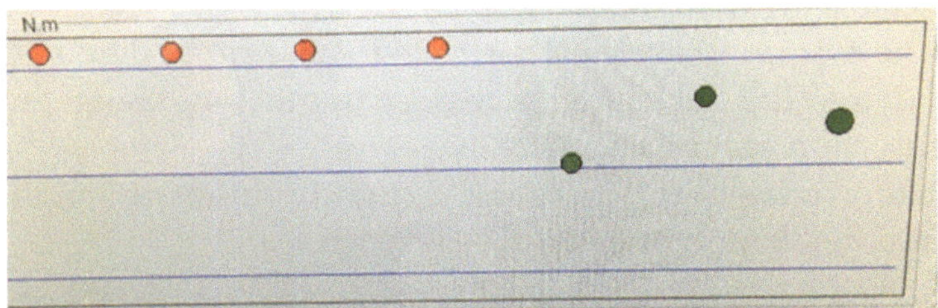

Calibration with proper part passes!

Torque wrenches - basics and calibration

Cheap / low quality

Don't use low quality tools.

Most of the first aid injury cases reported in the workplace such as cuts, blunt, abrasions, amputations, pinches or caught in between, etc. are due to injuries from hand tools. Improper use of hand tools has sometimes resulted in serious finger or eye injuries and can be responsible for permanent disabilities.

Ugga Dugga

Yes, ugga dugga made it (with a smile) into a book that is meant to provide serious information only.

But, in fact, ugga dugga is still well known in some places and still in use in some places in the US.
It's a remainer of a long gone time of easy to repair or easy maintenance pick-up trucks . Easy describes the technology used, not the repair work itself. So "ugga dugga" completes the torque subject in this book, knowing modern requirements don't allow such performance anymore

Unit of Measurement regarding time and usually torque. Often found in automotive stores, mechanic's garages, and redneck tractor pulls.

Ugga duggas can be counted infinitely, but it is commonly understood that 5 Ugga duggas is the maximum for torque, while three Ugga duggas is sufficient to hold most projects. To count an Ugga Dugga one must be wearing a ball cap backwards, have at least one busted knuckle, a beer in reach. Then as you squeeze the trigger of your impact gun, you count off like so, "One Ugga Dugga, Two Ugga Dugga, Three Ugga Dugga..etc"
Don't forget the southern, redneck drawl or to spit the juice from your oral tobacco products, or else your Ugga Duggas will not be sufficient.

https://www.urbandictionary.com/

Impact wrenches / nut runners

This chapter actually does not belong in this book – nut runners do not meet the requirements that a measuring instrument must have as explained in the chapter "Calibration".

Torque wrench owners usually shall look for an accredited <u>calibration</u> laboratory.

But, however, in some cases regarding torque wrenches and always when nut runners need to be covered, some testing might be applicable, too.

In fact, more and more <u>testing</u> laboratories show up, accredited according to ISO 17025:2017, too.

Machine capability testing

These tests are known as MSA "Machine capability testing", "Machine capability study" and similar.

The term "machine capability" comes from production engineering / process engineering and has nothing to do with calibration - (initially)!

In the area of industrial screw connections, however, requests for implementation are also made to calibration service providers (e.g. for torque or angle of rotation).

The term "machine capability" refers to a process that examines and can show whether a machine can produce with sufficient security against errors.

The automotive industry often requires proof of an MFU and the achievement or exceeding of specified parameters. These key values are the two indices / key figures C_m and C_{mK}.

The procedure of those tests is doing repeating tests und using mathematics such as standard deviation and variance to get these key figures C_m and C_{mK}.

These must meet a predefined value.

Machine capability studies are carried out over a short period of time. This essentially means that the machine and the method are included.
Influences of different materials, operators or environmental conditions <u>are not taken into account</u>. One also speaks of short-term capability examinations. This also results in the generally higher demands on the machine capability parameters.
The successful validation of the machine capability for all machines involved in the overall product is a fundamental basis for guaranteeing a high yield. It is also necessary for long-term monitoring of the machine or the process with the help of process capability studies. The results of machine capability studies are shown with the coefficients C_m and C_{mk}.
Attention, in English-speaking countries short-term skills are sometimes also given with C_p and C_{pk}.

VDI/VDE 2649:

VDI/VDE 2649 complete hadline is "Rotary tools for bolted connection - Guideline for comparative power-measurements of hydraulic impulse tools"
This guideline serves exclusively for the comparative performance measurement of impulse tools under precisely defined basic conditions. The measuring principle is based on the comparison of the preload force generation of a continuously rotating tool with an impulse tool under the same general conditions.

With the aid of this method, it is possible to determine which comparable torque an impulse wrench applies to a defined joint (test fixture). The corresponding torque of the impulse wrench is calculated via the generated preload force.
Therefore, it is essential that the test fixture remains stable in the relevant properties during the test.
A directly traceable measurement of the torque
for impulse wrenches is not possible.

Bibliography

Bibliography

Source literature / Sources used and cited:

BIPM
International vocabulary of metrology – Basic and general concepts and associated terms (VIM)
3rd edition, 2008 version with minor corrections

NIST, National Institute of weights and measures
The International System of Units (SI)

ISO 9001:2015
Quality management systems – Requirements

ISO 10012:2004-03
"Measurement management systems - Requirements for measurement processes and measuring equipment "

ISO / IEC 17025:2018
" General requirements for the competence of testing and calibration laboratories"

ISO 19011:2011-12
" Guidelines for auditing management systems"

ISO 6789-1:2017
"Assembly tools for screws and nuts – Hand torque tools – Part 1: Requirements and methods for design conformance testing and quality conformance testing: minimum requirements for declaration of conformance"

Bibliography

ISO 6789-2:2017
"Assembly tools for screws and nuts — Hand torque tools — Part 2: Requirements for calibration and determination of measurement uncertainty"

Bernd Pesch
"Messunsicherheit: Basiswissen für Einsteiger und Anwender"
Books on Demand. ISBN

Peter Jäger
„Calibration compendium"
Books on Demand, ISBN 9783750436039